T0103794

A PEEP
INTO VOID

A PEEP
INTO VOID

DURGATOSH PANDEY

PARTRIDGE

To order additional copies of this book, contact
Partridge India
000 800 10062 62
orders.india@partridgepublishing.com

www.partridgepublishing.com/india

Contents

Dedicated to my late father,
who taught me how to think

Acknowledgement

This book is a result of an idea that seeded in me during my teen years. That seed was nurtured by so many that it would not be possible to name and thank all of them. My father, Late Sudhakar Pandey and my mother Krishna Pandey provided me with an environment at home and an education in school that stimulated me to think out of the box. While brooding on the topic and writing the book, my wife Rambha has always been a pillar of strength. Thank you so much for your patience. Thanks also to my wonderful children, Pranav and Swasti, for being a source of happiness and encouragement at all times.

I must thank my brothers, Ashutosh and Paritosh, with whom I have shared my thoughts since childhood and it was they who suggested that the thoughts be compiled in the form of a book. I also remember and thank Prof Ajit Chaturvedi of the Indian Institute of Technology (Kanpur), my sister-in-law Archana Mishra, and my friends, especially Rajkumar and Priyaranjan for reading through the draft and their critical comments. My special thanks goes to my student, Dr Mahesh Sultania, for all the art-work, without which the book would be bland and tasteless.

Learning is a continuous process, not just in an individual, but also in the entire mankind throughout its evolution across centuries. There have been many scientists, philosophers, artists, teachers, seers, and others who have contributed to this process of learning. My thanks goes to each one of them who, through their efforts, have enabled the next generation to look a little further than the previous generation has seen.

Preface

"There is a voice that doesn't use words. Listen."

■ Rumi (a mystic poet)

This is the voice of the stars and the sun, planets and the moons, and also of the electrons and photons. They speak through their motion, they speak to one another through the attraction of gravity, they speak through their regularity, and they also speak through their randomness. Scientists listen to this voice of silence, and try to make sense by recognizing patterns and discovering laws that explain these patterns. This is also the voice that the mystics call as the voice of the soul. Through this voice, the consciousness within me communicates with the consciousness in you, and also connects with the universe outside. It is this voice that generates questions within a curious mind, and the answers to such questions can be sought by listening carefully to the same voice. Children have a remarkable faculty of curiosity, but unfortunately this curiosity and the keenness of listening to such voice fade as we grow up.

When I was a kid studying in a boarding school, we had some excellent teachers who left a deep imprint on many of us. One of them was Mr Ashok Kanti Ghosh, our Geography teacher, a very jovial and yet a strict person. Once he arrived to our class and narrated an incident: "You know, I was walking through the A-team playground when I suddenly noticed two snakes. The first snake was biting and eating the tail of the second one, and the second snake was in turn biting the tail of the first one and eating it too. The two snakes had grabbed each other's tails with their mouths and were eating each other. Suddenly I saw both of them disappear; they had eaten each other completely."

The whole class and Ashok da (as we used to call him fondly) had a hearty laugh. The anecdote was meant as a joke, and it was taken as a joke. Most of us forgot about it, but it remained in my subconscious mind, hidden within its multiple layers, only to resurface later when I started a deeper contemplation about the nature of reality and the origin of the universe. The Big-Bang theory of the origin of the universe could not explain the nature or the origin of the extremely dense plasma of matter-energy that expanded into the universe that we see now. It could also not explain the origin of entities like space and time. While meditating on these problems, Ashok da's anecdote flashed in my conscious mind, awakened from the deep recesses of my subconscious memory. I tried to mentally visualize the incident of the two snakes eating each other and disappearing into thin air. And then I tried to visualize it backwards as we sometimes watch a video or a film while rewinding it back. Two snakes appear out of blue, from nothing! The two snakes eating each other and disappearing was stupid enough; the appearance of the two snakes out of nothing was crazier than anything I had imagined before.

But then, what about the Big Bang that supposedly was the primordial explosion (if one may loosely call it) that created the universe? This was far crazier than the craziest thing that can be imagined by anyone. But now it made sense. Ashok da's anecdote, played backwards, giving rise to two snakes out of nothing, is qualitatively the same as the origin of the universe through the "Big Bang,", although the two differ in their scales and magnitudes. What banged in the "Big Bang" was nothingness or zero. Big Bang should then be the explosion of zero into the manifest universe. Nothingness or void should be the origin of the universe with all its entities and attributes. Space, time, matter, energy, consciousness, and all the stuff that we can imagine in this universe must be related to one another in such a way that their sum total is zero. This was the initial inspiration behind writing this book. But as time went by and my thought process evolved a bit more, I could notice some gaps in my hypothesis. The search for truth cannot be an isolated study of one of the aspects of the truth. All its aspects are inter-related. Thus in my own amateur way, I explored the mysterious nature of zero and the infinitesimals, the nature of light, theory of relativity, quantum theory, the discussions about consciousness in the ancient texts of the Upanishads, and so on. In the centre of all was zero. This reinterpretation of zero, in my view, explains the nature of reality and the universe.

The range of human knowledge is already vast, and yet it is only a grain of sand in the vast shore of the ocean of knowledge yet to be explored. In my exploration of the known and the unknown, some errors and inaccuracies might have inadvertently crept in despite great care on my part, and I must apologise in advance for any such slip. I am just a humble cancer surgeon. I confess that I am not an expert in physics or mathematics, and I have studied these subjects formally only till the 12th grade. This book is only an expression of an idea born out of an immense curiosity about the true nature of reality; the idea that has been within me for nearly two decades now. If the originality of this idea can be appreciated by the readers, my purpose would be solved.

1
Introduction

"Though my soul may set in darkness,
it will rise to perfect light.
I have loved the stars too fondly
to be fearful of the night."

■ Sarah Williams (from her poem
"The Old Astronomer to His Pupil")

What is the nature of reality? It is the most fundamental of all questions that has since millennia, haunted the minds of the men and women of science and religion, philosophers and mystics; and have also agitated or challenged many ordinary and lay persons like me time and again. This question, in its various forms and modifications, has also been asked by children with their inherent curiosity. Unable to correctly answer, most parents and teachers have either ignored the question, changed the topic, or worse, rebuked the kids for asking "silly" questions. The most damaging consequence of growing up is the loss of curiosity that is so inherent in a child.

We live in this world, a planet called earth. Earth happens to be one of the eight planets in the solar system that revolve around the sun. Our solar system is, in turn, a miniscule component of the Milky Way, the name given to the galaxy that we inhabit. The Milky Way itself contains numerous stars, our sun being just one of them. There are a vast number of other galaxies with innumerable stars and planets in each one of them. The universe is the sum total of whatever exists. It is difficult to imagine the scale of the vastness of the universe. Our nearest star (other than the sun) is so far away that its light takes about four years to reach us. The expanse of the universe is so vast that the standard units of distance become inadequate to describe the scale of distance between stars and galaxies. Such distances are often measured in light-years; one light-year being the distance that light would cover in one year. Light, as we know, travels at a speed of 300,000 kilometers per second. This speed would mean that light would cover 1,800,000 kilometers in a minute, 108,000,000 kilometers in an hour, 2,592,000,000 kilometers in a day, and 946,080,000,000 kilometers in a year. This is the distance of a light-year: 946 billion and 80 million kilometers. Our galaxy, the Milky-Way is estimated to have a diameter of about 100-120 thousand light-years. To comprehend how vast these distances are, if we were to reduce the scale and compress the diameter of the milky-way to 100 meters, the solar system would be no more than 1 millimeter in size, and the nearest star would be about 4.2 mm away. We shall do well to remember that inter-galactic distances would deal with millions of light-years. The diameter of the observable universe is estimated to be about 93 billion light-years.

It is not just space that is so huge, the vastness of time is also mind-boggling. The average life-span of a human being is 70 years. Jesus Christ was born some 2015 years ago; Buddha was born about 2600 years ago; the earliest human civilization dates to about ten to twelve thousand years ago. Human beings (Homo sapiens) evolved from the earlier primates about 170,000 years ago. Life arose on earth about 3.7 billion years ago. The solar system and our planet earth came into existence 4.56 billion years ago; the age of our Sun is around 4.6 billion years. Our galaxy, the Milky-Way was formed some 11 billion years ago, nearly 3 billion years after the so-called Big-Bang that is supposed to have given rise to the universe. The age of the universe (according to the standard Big-Bang model) is estimated to about 13.8 billion years. To put such vastness of time in perspective, Carl Sagan popularized the concept of "cosmic calendar" in his book "The Dragon of Edens." If the entire history of the universe is condensed to 1-year duration and we assume the big-bang to have occurred at the beginning of January 1 and the current time as the end of December 31 at midnight, our galaxy (the Milky Way) was formed on 15 March, Sun was formed on 31 August, and the earth came into being on 16 September. The first life in the form of unicellular prokaryotic organism appeared on 21 September, first multicellular life was formed on 5 December, dinosaurs roamed on the earth between 25 and 30 December, and the entire human history is just over an hour old. Agriculture began 28 seconds ago, Buddha and Christ appeared 6 seconds and 5 seconds ago respectively, the European renaissance happened 2 seconds back. The last one second has seen the industrial revolution, colonialization and decolonialization, American and French revolutions, the two world wars, and man's conquest of space and moon.

Consider now the matter in the universe. An average human weighs about 70 Kg, an elephant weighs about 5000 Kg, and the blue whale (largest animal on earth) is about 150,000 Kg. The mass of the earth itself is 5.9722×10^{29} Kg. There are seven more planets in the solar system other than the earth. Yet, the sun comprises about 99.86% of the total mass of the solar system. The mass of the Milky-Way is estimated to be around 6-7 X 10^{11} times the mass of the sun. The total mass of the observable universe is estimated to be 3.35 X 10^{54} Kg (based on calculations from its estimated volume and mean density). Of this, less than 10% can be accounted for by the stars, the rest being made up of dark matter and dark energy (so named because the scientists are still in the dark about their nature).

In this vast stretch of space and time, we humans are extremely tiny and insignificant entities. Or, are we somewhat more than that? It seems remarkable that we, as conscious and intelligent beings, have the courage to enquire into the grand designs of the universe. It is not surprising at all that we are not able to fathom many of the secrets of Nature; after all, the universe would not really care whether we existed or not. Rather, it is truly extraordinary that we are able to explain quite a few things about the underlying harmony and order of Nature. By the methods of science and some imagination, we are able to lay down the laws that hold good for many of the phenomena in this universe. It is a tribute to the curiosity and intelligence of mankind that we know whatever little we know about the working of the universe. This however has not been enough to quench the curiosity of man. Man has the audacity to enquire into not just the working ("how") of the universe, but also into the "why" of the universe: why does it exist, what is its origin, what is the nature of reality, and so on.

Since time immemorial, man has gazed up towards the sky, and has wondered. He has looked towards the sun, the moon, the stars, the constellations, and has tried to find patterns in their movements. The regular patterns of the day followed by night, and that of the seasons that appeared at regular intervals were clues to early man that the working of the world is not random, but can be explained by a set of defined laws. He was successful to a great extent in predicting and explaining the phenomena of eclipses, he was able to chart out the path of many of the stars and constellations. Astronomy evolved from a set of observations to a set of laws underlying the patterns in the universe. Physics defined and further refined these laws, ably assisted by other sciences. Mathematics was discovered, initially as a tool for counting and transactions between men, and to understand different geometrical shapes and figures found in nature. It was soon extended to the observations of astronomy to provide a uniform language for the laws of physics. Today, physics cannot be separated from mathematics.

"Where did it all come from?" This has been a question that has repeatedly troubled the intellect of mankind. Did the universe have an origin, or has it always existed? Why does the universe exist? These are questions that are still mysterious. Whether or not we will ever be able to find satisfactory answers to these fundamental questions is important, but what is more important is that we must continue to probe into these questions. They may not be of immediate or future benefit for mankind. These questions may not be relevant to our daily life; these may not help us in building machines that would make our lives simpler and richer. They may not be relevant to the material good of mankind. Or, who knows? Maybe, answers to such questions may be sources of other discoveries that may directly impact our material lives. Even if they do not, knowledge for its own sake is no less important. The pursuit of knowledge for the sake of knowledge itself is the purest of all intellectual pursuits. These questions into the "why" of the universe are definitely relevant in the domain of intellectual curiosity and growth.

You would have noticed that I have divided the questions about the universe between the "how" and "why" of the universe. Science generally deals with the questions of "how" because it is based on observations that can be reliably reproduced. Theoretical physics and mathematics have attempted to answer some of the questions relating to the "why", but even these must conform to the observational data, and thus become limited in their scope. Hence these fundamental questions have gone into the domain of metaphysics and religion, which do not necessarily need confirmation by observations. These disciplines of metaphysics and religion have their own shortcomings however. These rely on the existence of a Supreme Being, God, who is beyond comprehension by human mind. All such questions about the origin of universe and the nature of reality would finally have an answer pointing to the will of the almighty God. Unsatisfied by such answers offered by religion, these questions were also taken up by Philosophy, which literally means love for knowledge. The free-flowing thoughts of Socrates, Plato and Aristotle in the west and the philosophical discourses of the Upanishads and the Bhagwad-Geeta in the Indian tradition have also dealt with these questions in their own ways. The tradition of free flow of logical thoughts that characterized philosophy of the past was a fountain for many a scientific progress. Today, the discipline of philosophy has also been bound into several compartments, and it does not generate much discussion on the fundamental questions about the existence of the universe.

It is precisely these questions of "why" that this book attempts to address. I suppose this is the real quest of truth. In this quest, I have taken liberty to draw from the disciplines of science, philosophy as well as spirituality. Many of the thoughts that I have expressed are simply extensions of the already-existing knowledge. On some of these that I can claim as original, I may not be able to give evidences based on observations and on the exact methods of science. I would be glad if some physicists and mathematicians can expand on these thoughts to give a clearer picture of the universe. Only then we can know, as Einstein said, the mind of God.

2
Limitations of
our tools of observation:
The blind men and the elephant

"The first step towards knowledge is to know that we are ignorant."

■ Richard Cecil

Figure 1: Blind men and the elephant: Each of the
blind men is wise. But their perception of the reality
of an elephant is often erroneous and at best, incomplete.

We are all familiar with the story of the blind men and the elephant (*Figure 1*). Each of the blind men is wise. But because they cannot see the elephant as a whole, their descriptions of the animal are incomplete and often contradictory. One of them describes the legs of the elephant, the second describes its tusk, the third its trunk, the fourth its tail, and so on. Each one of them is correct in a limited sense, but their descriptions of the elephant are inadequate, if not incorrect. If these blind men co-operate with each other, they can make a very good description of the entire elephant, but they would still be unaware of the aspects of the elephant that are related to seeing the animal as a whole. We can understand this story because we can see and thus are aware of the limitations of these wise but blind men. These men are not limited in the sensory perceptions that they possess, nor are they limited in their intellectual capacity of reasoning and deduction. They simply lack the perception of sight, and this prevents them from being able to know what it is to see the elephant.

In science, all our theories are based on reasoning. Ultimately, they are based on some observations. These observations are then processed by our minds and machines that we have constructed. The sources of the raw materials for all scientific data and theories are observation and reasoning. Observation is the function of our sense organs and reasoning is the function of our mind and intellect. We have sensory perception for sight, hearing, touch, smell and taste. The machines and computers are just the extension of our sensory and mental capabilities. Thus, we have cameras and videos (extension of vision), radios and transistors and audio CD (extension of hearing), calculators and computers (extension of mental computing) and so on.

Most of us consider what we see as what really is true. The evidence of an eye-witness is a clinching evidence in many legal proceedings. Many scientific theories have been developed based on observations that are predominantly visual, other theories need observations to verify them. But is the observation by the eyes the reality of what we actually see? I am not going into the phenomena of illusions and magic and hypnosis here. My problem is simple: can we believe what we see as the reality of the object seen? Let us take the example of a table. We see a structure that appears solid to us and has a particular shape. If we look closely, we would see some rough edges, and irregularities in its otherwise smooth surface. Look with a powerful magnifying lens or a microscope, we would find several minute pores in the otherwise solid appearing wooden table. But is this what the table actually is?

If we go even deeper, the table is made up of atoms. The atoms in turn are composed of protons and neutrons that are confined in a nucleus, and electrons that whiz around the nucleus. These particles (protons, neutrons and electrons) are quantum particles that have highly unusual properties of dual nature (particle and wave), uncertainty and probabilities. But we shall not go into the intricacies of quantum particles as of now; we shall be discussing some of the aspects of quantum theory later in this book. Nonetheless, even the conventional view of the structure of an atom reveals something very strange. The nucleus (consisting of the protons and neutrons) comprise of almost the entire mass of the atoms (electrons are several degrees of magnitude smaller and lighter than the protons and neutrons); and yet, the nucleus occupies a miniscule part of the size of the atom. In fact, the nucleus occupies only about 1/100,000 of the space within an atom. The vast majority of the size of an atom is empty space. So the table that is composed of billions and trillions of such atoms must also be mostly an empty hollow structure. Regardless of the solid appearance of a table or chair, almost the entire structure (99.999% to be precise) of these "solid" objects is actually empty hollow space. The reality of the table is closer to being empty rather than being solid. And yet we rely on the visual perception as an accurate description of reality! Such an illusory perception of sight is true of whatever we see.

Why is this so? Why is it that even though 99.999% of the table actually is empty space, we see it as a solid structure? The answer would lie in the property of light that is absolutely essential to any visual perception, and also to the mechanism of sight. What we see as a table is actually the result of a series of steps that start from the light waves being reflected from the table to the retina of our eyes, excitation of the optic nerve endings within the retina, several chemical and electric phenomena that are transmitted from retina to the optic region of our brains through the complex optic pathway, and finally the recognition of the electrochemical information as a visual impression of the table. All these steps are filters between the actual table and what we perceive it to look like. Let us begin with the light waves. The visible light is a narrow band (380 to 760 nanometer wavelength) of the wide spectrum of electromagnetic waves; it is this narrow band that is responsible for the sense of sight. The electromagnetic waves of this spectrum have just the optimal energy able to cause excitation and conformational change of the molecules in the retina that leads to the electrochemical cascade through the optic nerves to the brain. Smaller wavelengths (ultraviolet rays) have much greater energy and cause damage to the retina; larger wavelengths (infrared rays) have much less energy to excite the molecules in the retina, and are thus unable to cause the electrochemical reactions within it. Now, the size of the atom is in the order of 10^{-10} meters ($1/10^{th}$ of a nanometer), and the size of the nucleus of an atom is in the order of 10^{-15} meters ($1/10,00,000^{th}$ of a nanometer). When the rays of light (of the visible spectrum) strikes the table, its wavelength is several orders of magnitude greater than the size of the atom and its nucleus. Because of the greater wavelength of the visible light, it cannot pass through the empty space of the atom; it must strike the nucleus in its path and is either absorbed, scattered or reflected. The reflected light from the table then would reach the retina where the visual cascade begins. So the reason why we don't see empty space in the table is not because the empty space does not exist, but because the light that is essential for us to see the table cannot pass through the empty space in the atom because of its wavelength that is several orders of magnitude larger than the atom itself. Electromagnetic waves of extremely short wavelengths (certain gamma rays) might be able to pass through the empty space within the atom, but these gamma rays cannot be perceived by our eyes as sight.

This is the limitation of the visible light, and this is not the only limitation of the entire mechanism of seeing. Visible light itself is the first filter between the reality of the table and what we actually see as the table. The next filter is the retina in our eyes where an inverted image of the table is formed after the reflected light passes through the cornea and the lens of the eyes, and is focused on the retina. Several nerve endings in the retina get excited by the photons of light that strike the retina. The inverted image of the table is then transformed into precise electrochemical reactions. This is the next filter. This electrochemical message is transmitted through the nerve fibers: one fibre transmits to the next until the information reaches the brain. Within the brain, the electrochemical information must be translated again into the visual picture of the table.

Just as the sense of sight is limited by these filters between the actual table and our visual impression of the table, similar arguments can be made regarding the other senses of hearing smell, taste and touch. We see therefore, that there are multiple filters between the reality of an object and our sensory perception of that object.

Science strives to explain all the phenomena in the universe by the scientific methods which rely on observations (however indirect and imperfect they may be) and their mental processing. We have seen how imperfect our sensory perceptions are in understanding the actual reality of the object being observed. Our understanding of the world is mostly through these imperfect sensory and mental faculties that we have. We know of the sensory and mental faculty that we are endowed with, but it is quite likely that we do not possess all the possible sensory perceptions or intellectual ability. We would never know about such sensory, mental and intellectual endowments that we probably do not have. I think it would be highly arrogant for us to think that human beings are endowed with all the possible sensory and mental faculties and these are enough to describe the wide-ranging phenomena in the universe. I shall illustrate this with some more examples. Human beings have the capability to hear the sounds that have frequencies in the range of 20 to 20,000 hertz (hearing range for humans). The hearing range of dogs is 50 to 46,000 hertz, and that of a bat is 3000 to 120,000 hertz. Bats are able to hear sounds in the ultrasonic range. The perception of smell of dogs is far stronger than that of human beings; this extraordinary perception of smell coupled with their loyalty has led to their use in crime investigations, detection of bombs, and other forensic applications. Many fishes, sharks, dolphins and the marine animals have the ability to sense the electric field, and utilize it for the search of prey. The migratory birds have the unique ability to detect the magnetic field of the earth and use it for their navigation across large distances in particular seasons and then come back to their original habitation. These sensory perceptions are either much weaker or absent in human beings. There may be other kinds of perceptions that we are not even aware of.

This brings us back to the story of the blind men and the elephant. Imagine a world in which everybody is blind. Nobody would be aware of sight as a sensory perception. Of course, they would have their own way to perceive and interpret things; they would have their own scientific theories about all their observations and various phenomena of the universe. But the fact remains that they cannot see. Now, imagine a situation where in such a world, as a result of some kind of mutation, someone is born with eyes. He can see things; to him, all the descriptive (non-visual) facts about the elephant hold no meaning because he can see the whole elephant. But the tragedy is that nobody would believe him because his experience cannot be replicated by others; a necessary requirement for scientific theories. This man with eyes would be regarded as unscientific and possibly mad. He would perhaps be treated as a threat to the society, a danger to the existing foundations of science, and also to the religious beliefs. Such an 'enlightened' person may be placed in solitary confinement, put in a mental asylum, or possibly be killed to "protect the society from this dangerous individual." Some may also revere him as a "Guru" or a founder of a new religion and faith. But nobody would actually understand him.

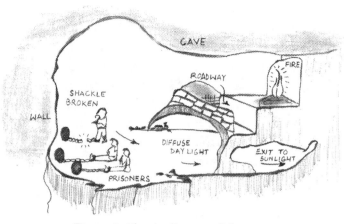

Figure 2: Plato's allegory of the cave:
The captives in the cave can only see the shadows
of the objects moving behind them.

This concept also finds an echo with Plato's "Allegory of Caves". I shall describe this with some modifications to further elucidate my point (*Figure 2*). In this thought experiment by the great ancient Greek philosopher Plato, let us consider that several men are trapped in a cave with their face towards the wall of the cave from their early childhood. Their movements are constrained including the movement of their heads so that the only direction in which they can see is towards the wall. There is a flame (source of light) behind their back and there are objects moving between the prisoners and the flame. The prisoners perceive the objects and their movements as shadows on the wall. These prisoners, who have only been exposed to such shadows and not the actual reality, perceive these shadows as real. They may develop scientific theories on the movement of these objects and the smartest among them is considered to be one who can predict the appearance and movement of these shadows with the greatest accuracy. Now, imagine a situation in which the shackles of one of these prisoners are opened and he is free to move. When he turns around, he can see the objects and the flame. Of course, he will be blinded temporarily by the brightness of the flame but he will acclimatize gradually. He may come out of the cave and see the world around him and his perceptions will change dramatically. He will see sun as the source of light, much greater than the flame in the cave; a world outside with so many fascinating animals, plants, inanimate objects and animate beings that he could not have imagined inside the cave. He starts pitying the prisoners of the cave and returns to the cave in order to free them from their ignorance. He tries to teach them about the actual nature of reality as opposed to that of the reflected shadows, about the wonderful world outside the cave. But he is unable to remove the chains and shackles with which these prisoners have been bound. How will the prisoners react? Perhaps, they will consider him as a fool, a madman, a threat to their peaceful society, a challenge to their existing scientific methods, and a blasphemy against their conventional wisdom. Some of these prisoners may put him on an exalted position of a guru or a god man. Finally, he may meet the fate of Christ or Socrates by being crucified or poisoned to "protect the society from such radical views."

These thought experiments about the blind world and that of Plato's "Allegory of the cave" clearly show that it is wrong to rely completely on observations and our tools of observations to understand reality. The scientific method, which gives too much importance to observational data and reproducibility of observations, may not be enough to remove the veils from the nature of reality. I am not belittling the value of the conventional scientific methods. These methods have helped explain numerous phenomena of nature and have helped us understand the universe in an objective way. I however believe that they are not adequate for a full understanding of the secrets of the universe; we need other methods too.

Now let us consider the human species, the *Homo sapiens*. It is quite possible, rather extremely likely that there may be some sensory or mental faculty that we do not have. If a person is either born with or somehow acquires such a perception, the scientific community would tend to ignore it or worse, perhaps ridicule it. Such a person would find it extremely difficult to explain his experiences directly to others. Perhaps, Buddha and Christ were examples of such people. They had experiences that we possibly cannot have. Because it was not possible for them to explain their experiences directly, they used words that we could understand. But we all know that it is one thing to describe an elephant to somebody who cannot see; it is a completely different experience to see the elephant! Any amount of verbal description cannot match the act of seeing the elephant; similarly any degree of skill in any language cannot match the direct perception of something that is indescribable by language. No wonder that we do not understand them completely and are always in conflict. We all are like the blind men who are looking at the reality from our own individual limited perspectives. Just as one blind person describes the trunk of the elephant, another describes the tusk, still another describes the limbs; similarly some of us attempt to describe reality through science, others through philosophy, still others through literature, art, music, religion, faith and so on.

Let us all shed this misplaced arrogance of our species and accept that there are many things that we would be unable to explain because of our inherent limitations. Let us also be open to a completely different approach to a particular problem. This is the only way to scientific progress.

3
Compartmentalization of knowledge

"Science is the knowledge of secondary causes, of the created details; wisdom is the knowledge of primary causes, of the Uncreated Principle."

■ Dr S Radhakrishnan

Ever since life arose in this world, the living organisms have been adjusting with their surroundings; sometimes as a matter of necessity for their survival, at other times, to mould the surroundings to their own benefit. Man in particular has been endowed with the qualities of intelligence and discrimination through which he has tried to understand nature and sometimes control it, for the practical benefits of living as well as for the joy of pure knowledge. In the ancient times when mankind was just taking infant steps on this planet, the primary issue at hand was survival. The most important questions were related to food, shelter and safety. The quest for knowledge was towards these ends. The discovery of fire, fashioning of primitive weapons, beginning of agriculture, building of houses, invention of wheels -- all these were products of human endeavour that made his life more secure and comfortable. As he continued to evolve, the spirit of enquiry developed further and man began thinking about the questions of ethics and justice, social order and politics, disease and medicine, consciousness and soul, religion and morality, truth and beauty, and above all about the mystery of existence.

Knowledge then was one organic whole. It was not split into different compartments as we see today. Thinkers and philosophers were truly free in their thoughts and work. We find people like Pythagoras, Socrates, Plato and Aristotle in the west who contemplated and wrote on such a wide range that is almost impossible to conceive now. They taught and wrote on mathematics, geometry, science, politics, ethics, justice, forms of government, and so on. In India, we find the magnificent compilation of the Vedas and Upanishads that deal with the real nature of consciousness and self, and also prescribe the correct way of living for individuals and societies. They also deal with origin of universe, mathematics and medicine. Aryabhatta invented zero as place value of numbers, and Brahmagupta formulated the rules governing zero that revolutionized mathematics. Baudhayana, another Indian mathematician independently described the theorem of right angled triangle that famously goes by the name of Pythagoras; he did so around 800 BC, about 300 years before Pythagoras. The developments in mathematics and geometry were closely followed by the progress in astronomy and astrology that were closely knit then. Knowledge of one discipline could act as a catalyst for the knowledge of another discipline, often in the same individual.

This trend continued till the advent of the modern age. Just about 500 years ago, we see the examples of geniuses like Leonardo da Vinci who was at once a painter, engineer, sculptor, architect, anatomist and thinker; and Isaac Newton who wrote as much in theology as in mathematics and science. Simultaneously, there were many who were not formally trained in science but had an amateurish interest that led them to make such discoveries that revolutionized science. For example, Gregor Johann Mendel was a priest but his keen observations and experiments in his garden led to the formulation of the laws of hereditary transmission of traits upon which the entire field of genetics is based. Erwin Schrodinger, one of the key physicists in the development of quantum theory, also wrote a book "What is Life" in which he described the concept of a complex molecule that carries the genetic information of living beings. Schrodinger's book was an inspiration for the later discovery of the double helical structure of DNA (deoxyribonucleic acid) by James Watson and Francis Crick. Throughout the ages, this intermingling of various diverse disciplines of arts and science has been a source of new thoughts and ideas.

In contrast, knowledge today has become unmanageably vast and diverse. There has been an almost irreversible trend of specialization and sub-specialization in each discipline of knowledge. In medicine, we had physicians and surgeons; now we have a whole gamut of sub-specializations in each branch of medicine and surgery: cardiology and cardiac surgery; neurology and neurosurgery; surgical, medical and radiation oncology; pediatrics and pediatric surgery; orthopaedics, gynecology, ophthalmology, otorhinolaryngology, geriatrics, endocrinology, and many other such subspecialities. In mathematics, we have people who specialize in statistics, field theory, set theory, and so on. In physics, we have particle physics, quantum mechanics, applied physics, optics and so on. There is an incentive to know more and more about less and less. This trend of specialization has its obvious merits; the specialist can focus his time, talent and energy to a small area of knowledge and become a real expert in that field. Such specialists can dive deep into their respective fields and bring out pearls of knowledge hidden from common people. Despite these virtues, excessive specialization has led to more and more emphasis on accumulation of facts and less and less on synthesis of those facts into holistic understanding. The specialist will often close his eyes to the other discipline, have a tubular vision of his own speciality, and thus will likely lose the broader perspective. Every branch of knowledge, more specifically every branch of science, has developed an edifice of technical terminologies and a unique language that can be understood only by the specialist of that particular branch. The specialist finds it difficult to communicate his understanding in common language and thus each branch of knowledge is understood, practiced and communicated within its own practitioners.

But is "reality" actually fragmented into different disciplines, or is it one organic whole that finds itself artificially divided into specialities? The development of specialization was meant to dig deeper into every area of knowledge so as to aid the wise persons and teachers to use their insights to develop fresh perspectives and solutions to unresolved problems. Instead it has led to development of different creeds of specialists who do not understand one another's languages and terminologies; neither are they intelligible to the common man. Not unlike different religions that are often at conflict based on the superficial rituals while forgetting that there is a common underlying truth, these different specialities are indifferent to one another forgetting that they are looking at a small aspect of reality that is actually one organic whole. Erwin Schrodinger, one of the eminent theoretical physicists who contributed greatly to the development of quantum mechanics, warns that if scientists "continue musing to each other in terms that are, at best, understood by a small group of close fellow travellers," then science "is bound to atrophy and ossify." It is not that the men and women of science do not understand

and appreciate the limitations of excessive specialization. Attempts at integration of physical and biological sciences have given birth to new areas of specializations like biophysics, biochemistry and biotechnology. The use of lasers in surgery and microscopes in pathology are examples of the application of optics in medicine. Such attempts at integration and unification of different streams of science, however limited they might be, are welcome, and we need many more such endeavours. They are supposed to build bridges between different disciplines; but they often get in the same old trap of becoming yet another new speciality with its own lingo and terminologies.

The responsibility of synthesizing facts, forming perspective, and translating knowledge into wisdom was, in the past, that of the philosophers. History has witnessed great philosophers in both the east and the west who were teachers and leaders of all sections of the society, from kings to paupers. Vyas, Nagarjuna and Shankar in India; Confucius and Lao Tzu in China; Socrates, Plato and Aristotle in Greece were shining examples of such eminent philosophers in the ancient times. Philosophy used to be the fountainhead of all branches of science and art; it was the advancing edge of all knowledge. Unfortunately today, philosophy has not been able to keep pace with the rapid advances in its daughter disciplines of science and art. Like a father who has been ignored and alienated by his children, philosophy has lost its confidence. Unable to assert itself among the specialities of science, it has taken refuge into a narrow corner. What a shame and letdown it has been! While philosophy in the past used science as a tool to aid in the synthetic interpretation of nature, today the philosophers have surrendered meekly to the process of fragmentation of the knowledge, have divided their own discipline of philosophy into artificial compartments and obviously cannot find courage to follow those glorious traditions of the past.

What has been the result of fragmentation and compartmentalization of knowledge? The consequences are far reaching. Too much emphasis has been placed on facts, too little on understanding. Unable to comprehend and scared of facing the whole, the specialists have busied themselves with fragments, thus becoming oblivious to the holistic understanding of nature. In the process of describing the details of the trees, the idea of forest has been given up. Perspective is lost, and wisdom, already an illusory concept, has been forgotten. Generalists and free thinkers are treated with contempt by the scientific specialists, to an extent that their methods of logical thinking and deductive analysis are called pseudo-science. Scientific method, with all its shortcomings, reigns supreme and unchallenged. Questions that are uncomfortable or do not yield to "scientific method" are censored. Science has contended itself to the questions of "how" rather than taking the challenges of the questions pertaining to "why."

The answer to the questions of "why" needs holistic approach to knowledge. It cries for the wisdom of philosophy, for the synthetic interpretation of nature, for logical thinking and for the courage to confront the established norms of "scientific methods." Just as philosophy needs science to prove concepts and collect facts before it can synthesize such concepts and facts into a wholesome knowledge, science and the methods of science too need philosophy to build bridges with other disciplines and have a wider perspective of reality. Francis Bacon put it succinctly: "What science needs is philosophy – the analysis of scientific method, and the co-ordination of scientific purposes and results; without this, any science must be superficial. For as no perfect view of a country can be taken from a flat; so it is impossible to discover the remote and deep parts of any science by standing upon the levels of the same science, or without ascending to a higher." Philosophy as a discipline and philosophers as the teachers and students of this discipline must rise to this challenge and embrace upon themselves this stupendous responsibility of co-ordination, synthesis and interpretation of various disciplines of art and science.

4
Critique of scientific method

"A theory can be proved by experiment; but no path leads from experiment to the birth of a theory."

■ Albert Einstein

Science has helped us to explore the vastness of the universe, peep into the nature of subatomic particles, understand the causes of natural phenomena (like earthquakes, volcanoes, solar eclipse, etc), build machines to make our lives comfortable, treat diseases, and so on. There is hardly any aspect of our lives that has remained untouched by science. But science is not just a modern development; it has been practiced by man since antiquity. In the ancient times, man discovered fire that could keep him warm and scare away the wild animals, invented tools for protection against animals and weapons to hunt them, invented wheels, started agriculture, and also made significant intellectual advances in astronomy and mathematics. As we discussed in the previous chapter, knowledge was one whole in those times and had not been fragmented into several specialities of science and art and philosophy. Thinkers during those times dealt with the entire spectrum of the known, the knowable, and the unknown. In many ways, this was an advantage, and in other ways, this was also a disadvantage. While the obvious disadvantage was a real possibility of reason being subjugated to faith, the advantage was the ability to look at the whole picture rather than having a selective and tubular vision, something that almost defines the specialists today. This approach to synthetic thought process gave rise to "Philosophy" which became the fountain-source of all of science and much of art.

The methods of science have been a subject of considerable debate over centuries. In the ancient times when philosophy dominated the intellectual discourse, intuition and logical reasoning were the two pillars of scientific enquiry. The earliest philosophers in India relied much upon the power of intuitive knowledge – the insight about something that comes in a flash of inspiration. In this way, Vedas and Upanishads were composed. Of course, behind this flash of inspirational insight lay a lot of hard work, deep meditation and thought over a particular problem. The early western philosophers relied much on logical reasoning. This method of logic led the philosophers to deduce one thing from a given fact or observation, and then to another, till a suitable answer was found to a particular problem. In this manner, Socrates, Plato and Aristotle dealt with problems of ethics, justice, politics, and science.

Modern scientific methods have evolved from the days of antiquity and have been carefully formulated so that the potential for errors, misrepresentation and misuse could be minimized. Observations and experiments became key elements of scientific method. Observation is not a new thing; from the infancy of mankind, man has been watching the sky with child-like wonder and curiosity, and has been trying to make sense of the movements of planets and stars. With the tool of observation, man has been able to find patterns in the change of seasons, natural phenomena like solar eclipses, and so on. Observations also led man to formulate several mathematical axioms; such as two parallel lines would never meet, there can be only one line that can be drawn through two given points, and so on.

Combining observations and deductive reasoning, Nicolaus Copernicus concluded that earth is not the centre of our solar system, and that it moves around the sun rather than the sun moving around the earth. It was contrary to the direct visual experience that sun rises in the east in the morning and sets in the west in the evenings every day, only to rise in the east yet again every next morning. This was sacrilege to the Christian belief in geocentric universe then. Copernicus died soon after his book about his model of the universe was published; this book was later banned by the church. Giordano Bruno went even further than the concept of the sun-centered solar system; he even proposed that the stars were just distant suns and that the universe is infinite with no celestial body as its center. For this and some more of his proclamations against the religious beliefs, Bruno was tried for heresy and was burnt at stake at Rome in the year 1600. However, despite all the obstacles, science marches forward with tentative steps, slowly but quite surely. It was the discovery of telescope by Dutch astronomers and its refinement by Galileo that revolutionized the art of observation and catapulted its importance in the methods of scientific enquiry. Telescope was a powerful means of astronomical observations and with this tool and his analytical skills, Galileo was convinced about the truth of earth revolving around the sun. His writings about heliocentrism (sun being the center of the solar system) infuriated the church. He was threatened by Pope and was forced to apologize and withdraw his position on this matter. He was imprisoned and spent the rest of his life in house arrest. Subsequent generations of scientists have contributed to the refinement of tools of observation and to the evolution of the modern scientific methods.

The Oxford English Dictionary defines Scientific method as "a method or procedure that has characterized natural science since the 17th century, consisting of systematic observation, measurement, and experiment, and the formulation, testing, and modification of hypotheses." Today, the scientific method has the following four components:

1. Making observations in nature and posing questions related to observations about a phenomenon.
2. Formulation of a hypothesis to explain the observations and the phenomena.
3. Use of the hypothesis to develop predictions about other phenomena that can be tested.
4. Gathering further data, making new observations and performing experiments to test the predictions by several independent experimenters and properly performed experiments.

If the experiments verify the hypothesis, it may be regarded as a theory or law of nature. If, on the other hand, the experiments are at variance with the hypothesis, the hypothesis should be rejected or modified. It must also be emphasized that reproducibility of the results by different experimenters is also an important characteristic of scientific method. If an experiment is not reproducible, it does not qualify to prove a hypothesis. A scientific theory is subject to falsification if new experimental or observational evidence incompatible to it is found. In such situations, a new theory may be proposed that better explains the given phenomena, or minor modifications are made to the existing theory to take care of the incompatibility. In fact, established theories can also be subsumed by more accurate theories. Ptolemy's theory of geocentric universe was given up when it was discovered that earth is just one of the planets in the solar system that revolves around the sun. Newton's laws of motion and gravity explained to a great degree of precision the planetary motions and these laws were considered sacred for centuries. However, they were subsumed by a more accurate theory of relativity proposed by Einstein that explained the minor inconsistencies of Newton's laws and also predicted other phenomena such as bending of light by gravity, that were subsequently observed.

While these scientific methods are extremely valuable in explaining the observed phenomena of nature and in making sense of several sets of observations, integrating them into a set of well-defined laws, there are likely to be questions and problems to which there could possibly be no experimental or observational evidence. In such situations, one must turn to other branches of knowledge like art and philosophy while searching for answer, and should not give up just because of the limitations of the "methods" of its discipline. After all, if we dig deeper into the literal meanings of the words "science" and "philosophy," they are almost identical. While "Science" is derived from Latin "scientia" which again is derived from "scire" meaning "to know" and "to discern," the word "Philosophy" stems from "philo" meaning "love for" and sophia" meaning "wisdom." Philosophy thus is the love for wisdom; science is the pursuit of knowledge. Both are complementary to each other; and it would be a tragedy if the practitioners of science were to see the discipline of philosophy as something that is irrational and unreliable.

While the scientific methods are useful in explaining the questions of "hows" in nature, they are often insufficient in explaining the questions pertaining to the "whys." We must, in such situations, take help from logical analysis, which is the heart and soul of the methods of philosophy. Let us understand this with the help of a few illustrations. We shall first see the application of scientific methods in some of the most important advances in modern science. Then we shall move on to other questions where these scientific methods are insufficient.

Newton's laws of motion

The three laws of motion that Isaac Newton described has been a standard part of all science textbooks taught in schools. The first law states that an object would continue to be in its state of rest or uniform motion in the same direction unless an external force acts on it. This is often called the law of inertia. Inertia of a body at rest is easy to understand; it obviously needs an external force to move something that is at rest. Inertia of motion seems contrary to our daily experience as a ball moving on the ground does come to stop. However, if we look at it with a little more attention, there is a force that stops the moving ball, and this force is the friction between the ball and the ground. If this force of friction is completely removed, the ball would indeed remain in perpetual motion.

The second law of motion is about the relation between force applied on a body and the acceleration that results from it. It states that acceleration of a body (that is the rate of change of its momentum) is proportional to the force applied on it. It is a scientific law explaining the obvious fact that it takes greater force to give the same acceleration to a heavier object than to a lighter one. Newton gave an exact mathematical relationship between force applied (F), mass of the object (M) and acceleration (A): $F = M \times A$ (force is mass times acceleration).

The third law of motion states that for every action there is an equal and opposite reaction. When you push a wall, you are also pushed back equally hard by the wall. The recoil of a rifle when a bullet is shot is an example of the third law. Propulsion of rocket relies on the third law of motion. The powerful engines of a rocket push down with a great force and reaction of that force propels the rocket upwards with an equal force.

All these three laws of motion were based on careful observation of the natural phenomena and experiments to verify them.

Newton's law of gravity

It was perhaps an apple falling on Newton's head that triggered his thoughts on gravitational attraction. Or, may be, the story is a myth that has been propagated through centuries. Whether the apple actually fell on his head or not, Newton wished to explain this attractive force between two bodies. He formulated the hypothesis that the gravitational force of attraction between any two bodies is directly proportional to the product of their masses and inversely proportional to the square of the distance between them.

$F = G\, m_1 m_2 / r^2$, where   F is the gravitational force,
G is the gravitational constant,
m_1 and m_2 are the masses of the bodies in question,
and r is the distance between the two bodies.

This hypothesis was subjected to several experimental tests involving several bodies, and was subsequently accepted to be a valid theory. Newton's law of gravitation explained that the force responsible for the apple to fall on the earth is also responsible for the moon to orbit around the earth. It could also explain the movements of planets around the sun. There was a small discrepancy between the prediction of Newton's law of gravity and the precession of Mercury's orbit. The orbit of Mercury appears to be an ellipse; however the point in this ellipse that is closest to the sun is not constant, rather it slowly moves around the sun: this movement of the orbit is called precession. Not just Mercury, all planets exhibit the property of precession. While the precession of all other planets of the Solar system could be explained using Newton's equations, Mercury seems to be an exception. Although this discrepancy was ignored for quite some time and alternative explanations were given, Einstein's theory of general relativity explained gravity in a better way and could also predict the precession of the orbit of Mercury quite accurately.

Einstein's theory of special relativity

Einstein found a contradiction between Maxwell's description of electromagnetic fields and classical mechanics as explained by Galileo and Newton. James Clerk Maxwell formulated the theory of electromagnetic waves and had shown that these waves propagate with a fixed speed 'c' (the speed of light), regardless of whether the frame of reference was stationary or moving. In addition to unifying electricity and magnetism, Maxwell also concluded that light is an electromagnetic wave propagating through the field according to electromagnetic laws. According to the previously held classical description of relativity, the physical laws of motion described by Galileo and Newton should remain unchanged if we pass from a stationary to a moving frame of reference. Thus, if you throw a ball up in a moving train, it will fall back to you in the same way as it would have come back to you if you had thrown it up while sitting in a garden. If a person on the platform fires a bullet in the same direction as that of a moving train, a passenger in the train will perceive the speed of the bullet to be less than what will be perceived by the person on the platform. Einstein applied the same logic to very high speeds. If we travel very rapidly with a velocity v in some direction, then the speed of light in the same direction ought to be reduced to less than c (c-v), and the speed of light in the opposite direction ought to be increased to more than c (c+v). However, Maxwell had derived a constant value of the speed of light in vacuum that should be applicable to all frames of reference. This was the contradiction that led Einstein to his theory of special relativity.

In a stroke of genius, Einstein proposed that there is no absolute inertial frame of reference and he completely turned around the prevalent notions about space and time. He said that our measurements of distance, mass and time are relative – they are not the same for all observers; they have different values in different reference frames. The speed of light remains constant in all reference frames. When an object, say a jet, moves at a very high speed (close to the speed of light), it would appear to be shorter (length contraction), heavier (its mass would increase), and its clock seems to run slower (time dilation) than an identical stationary object.

Einstein did not prove his hypothesis of special relativity, which to him was logical and natural. Rather, he made predictions based on these hypotheses, and these predictions have been confirmed to great precision later by several experiments. One of the most stunning verification of "time dilation" was the experiment that was published by JC Hafele and Richard E

Keating in the journal Science in 1972. In this experiment, atomic clocks were flown on commercial airlines around the world in both directions, and time elapsed on the airborne clocks were compared with the time elapsed on the earthbound clock. The clock flying eastward lost 59 nanoseconds and the clock flying westward gained 273 nanoseconds as compared to the earthbound clock. The "clock paradox" predicted by Einstein's special relativity was thus proved by experiments.

Einstein's theory of general relativity

The limitation of the theory of special relativity was that it applied to inertial frames of reference; i.e. to bodies moving at a constant velocity. It did not apply to accelerating bodies and thus could not account for gravity which is an omnipresent accelerating force. By this time, Minkowsky, who had been Einstein's mathematics teacher at polytechnic, had come up with a magnificent geometrical insight of a four-dimensional space-time (including the 3-dimensional space and the 4th dimension of time).

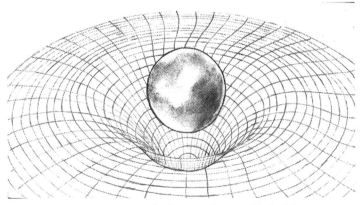

Figure 3: Einstein's theory of general relativity explains gravity as the alteration of the geometry of the space-time fabric by a massive object.

In another remarkable flash of insight, Einstein postulated that gravity is nothing but the effect of curvature that a body with mass induces on the 4-dimensional space-time fabric. Another body just follows the shortest path in this space-time curvature. It so happens that the shortest path of the earth in the space-time curvature induced by the mass of the sun is the earth's orbit around the sun, the shortest path of an apple falling from a tree is a straight line towards the centre of the earth. General relativity transformed the physics of gravity to geometry. Gravity was thus described by Einstein not as a force but as a result of change in geometry of space-time curvature induced by a massive object (*Figure 3*). It can be understood a little better by an analogy of a ball that is placed on a two-dimensional flat net. The mass of the ball would cause the two-dimensional flat net to have a curvature. If another smaller object is placed on this net, it follows the curvature induced by the more massive ball and is thus attracted to the ball. Similarly, a massive object would induce a curvature of the space-time (distortion of its four-dimensional structure) and attract other objects that follow this curvature.

While the mathematics underlying general relativity is quite complex and beyond the understanding of lay people like me, the beauty and elegance of the theory describing the interaction of space-time with matter and energy is breathtaking. A famous physicist John Wheeler described the general relativity as: "Mass and energy tell space and time how to curve, and space and time tell mass and energy how to move."

Again, Einstein did not attempt to prove his theory with observations. Instead he made predictions based on general relativity. He proposed that light should also bend following the bending of space-time curvature by massive objects like stars, and predicted that starlight passing just above the sun's surface should bend by 1/2000th of a degree; light passing twice as far from the sun's centre should bend half as much. Four years after Einstein published his theory of general relativity, Arthur Eddington confirmed this small bending of light by observations during a total solar eclipse.

The precession of planet Mercury, which could not be explained by Newton's law of gravity, could now be accurately explained using general relativity. Other applications of general relativity were gravitational lensing of light (used as a tool by astronomers) and GPS (Global Positioning System) satellite navigation system. In fact, if the effect of special and general relativity were not taken into consideration for the GPS, errors in global positioning would accumulate at the rate of about 10 kilometers each day. The whole system would be an utter failure for the purpose of navigation.

Quantum theory

We have seen so far that the experiments and observations as necessary requirements for a scientific method have been successful in the scientific developments described above. In fact, most of the earlier theories of classical physics were the results of careful observations and scrupulous experiments. Einstein changed the approach of modern science to some extent. His theories of special and general relativity were founded on logical and intelligent assumptions. He didn't care for giving any observational and experimental proofs for his theories of relativity; rather he thought them to be of secondary importance. The experimental and observational proofs of the predictions of his theory were later performed by other scientists, while he himself had no doubts about the veracity of his principles of relativity. It was in this context that Einstein had said, "A theory can be proved by experiment; but no path leads from experiment to the birth of a theory."

The scientific theories described so far are theories of classical physics. However, these were not enough to describe the behavior of submicroscopic world; the realms of atoms and subatomic particles. At those dimensions, the classical physics fails and one has to rely on apparently weird quantum theory. Despite its weirdness, quantum theory is one of the most important advancement in our pursuit of knowing the nature of reality. It appears to be a very disconcerting theory initially as it seems to destroy the beauty of the inherent symmetry of nature, brings uncertainty as a central theme, and uses probability to describe the nature of reality. Nothing seems real in this theory and the apparent reality is perceived as a result of the interaction of observer with the observed. It is no wonder that a great mind like Einstein could not come to terms with the probabilistic nature of quantum theory despite being one of the pioneers in laying the foundation of this theory. "God does not play dice...", he famously said.

The very foundation of observation and experiments as necessary requirements of the "scientific method" lay on the distinction between the observer and the observed. According to classical physics, there is an objective world that can be a subject of experiments and observations by an observer. The observer merely observes the world and makes attempts to uncover the secrets of nature. The universe will continue to behave in exactly the same way it does, whether or not there is an observer to observe it. In contrast, the quantum mechanics (more specifically the Copenhagen interpretation of quantum mechanics) rejects the objective reality of the quantum microworld. The distinction between the observer and observed is not just blurred but the observer becomes a necessary part of the whole system. As Werner Heisenberg, the discoverer of uncertainty principle, said, "… we have to remember that what we observe is not nature in itself but nature exposed to our method of questioning." Neils Bohr, one of the founding fathers of quantum theory succinctly described, "When searching for harmony in life one must never forget that in the drama of existence we are ourselves both actors and spectators."

The dual nature of light (as particle and wave) extended to other particles like electrons too. The uncertainty principle suggested that a particle (say an electron or a photon) does not possess an objectively defined position and momentum at the same time. How an electron moves or behaves is a superposition of an entire range of possibilities or potentia; it is the act of observation that reduces the infinite range of possibilities to one ("collapse of wave function," as termed in quantum mechanics). These concepts were truly revolutionary and contrary to the established way of science and scientific method. Yet, quantum theory has stood the test of time and despite the many uncomfortable premises and suppositions of this theory, it describes the working of atoms very well, even though in statistical terms. The "collapse of wave function" by the act of observation hints at the importance of the phenomenon of consciousness in the quantum world.

Thus we have now got the acknowledgement of the limitations of the scientific method from the very science that once laid these rigorous methods. The quantum world cannot be understood by observations alone, the theory of relativity was not based on observational data (its predictions were later successfully confirmed by experiments), and the phenomenon of consciousness defies all probing by the scientific method.

We shall come back to the quantum theory again and I shall attempt to explain its principles later, based on my interpretation of the nature of zero. We shall also come back later to the mysterious nature of consciousness in the light of the quantum theory.

5

The basic components of universe: matter, energy, space and time

"The nitrogen in our DNA, the calcium in our teeth, the iron in our blood, the carbon in our apple pies were made in the interiors of the collapsing stars. We are made of starstuff."

■ Carl Sagan

"Do not feel lonely; the entire universe is inside you."

■ Rumi

What comprises the universe? There are planets, stars, comets, meteors, galaxies, etc spread out in a vast expanse of space. These can be grouped together as "matter" that have mass and can be divided and subdivided into molecules, atoms, protons, neutrons, electrons and other subatomic particles. The "space" that contains the matter of the universe is another component. The universe is not static, it is changing with time. New stars and galaxies are born, older ones die or get transformed into nova, supernova, black-holes etc. "Time" then is another component that characterizes the universe; the entire drama of the universe is unfolded in the passage of time. The dynamic nature of the universe also means that matter would be created, destroyed, or would change its form. All this would require either release or utilization of energy. The matter that is in motion would also have kinetic energy associated with it. Thus, "energy" is another component of the universe. There are living beings including us humans who observe the phenomena in the universe, and use intelligence to explain and describe these phenomena. We are ultimately composed of matter; of several molecules that interact with each other to form cells, tissues and organs. Yet, "consciousness" in human beings and others that observe the phenomena of nature might be the fifth basic component. The nature of consciousness, however, is a matter of considerable controversy and debate, and may lead us to the realms of metaphysics. We have seen the limitations of science and its methods in explaining the nature of consciousness. Drifting our discussion to metaphysics is not the intent of this book and hence, we shall exclude the discussion about consciousness as of now and confine ourselves to the material aspect of the universe. We shall return to the topic of consciousness again in the last parts of the book.

The conventional way in which the scientific community describes the universe ultimately boils down to these few basic components. Regardless of the problems about the origin and evolution of the universe, the basic ingredients that would perhaps define the physical universe are space, time, matter and energy. All other things can be explained by these basic components and their interactions. For example, motion can be explained in terms of matter moving in time through space. Forces that attract or repel the particles or bodies (gravity, electromagnetic forces, weak and strong nuclear forces) can be explained in terms of interactions between the matters that constitute them.

What then is the nature of these basic components? Let us examine them in a greater detail, one at a time.

Matter

Matter is what we see all around us. The tables, chairs, houses, air, water, trees, moon, earth, sun, stars and everything we can imagine can be included within this term, "matter." All living beings including humans are also comprised of matter. For centuries, there had been a debate whether matter is continuous or discrete. Many believed that it is continuous, even when divided and further subdivided indefinitely. It was Democritus, a Greek mathematician and philosopher, who first declared matter to be discrete, and he named the smallest building block of matter as "atom," literally meaning uncuttable. Of course, there was no way of "seeing" or directly detecting atoms in those days or even in subsequent centuries, and the debate on the atomic nature of matter continued.

In the 19th century, two important observations pointed towards the atomic nature of matter. The first was that of the English botanist Robert Brown who observed under microscope, the random motion of pollen grains suspended in water. This random motion of the pollen grains, the "Brownian motion" could not be explained satisfactorily. Some scientists suggested that Brownian motion was a result of the collision of atoms, but a satisfactory theory could not be formulated. The second development was the analysis of thermodynamics by an Austrian physicist, Ludwig Boltzmann. He proposed that the reason matter could be heated and the heat transferred from an object at higher temperature to another object at lower temperature, was that matter was ultimately made up of atoms. Atoms of hotter objects move faster and have a greater kinetic energy than colder object. The transfer of heat is because of the transfer of energy from faster to slower atoms till the temperature equivalence is reached. Even though Boltzmann could very well explain the laws of thermodynamics using the concept of atoms comprising matter, the general scientific opinion wanted a more direct proof. This proof came from Einstein in the beginning of 20th century when he gave a detailed description of the Brownian motion and could predict the distance travelled by a pollen grain in a liquid medium based on the size of the grains, viscosity of the liquid medium, temperature of the liquid medium, and time. He explained the Brownian motion to be indeed because of collisions between the atoms. Finally, the debate was settled and the atomic nature of matter became widely accepted.

But, this was not enough. Atoms are not "uncuttable" as Democritus thought them to be; rather they are composed of a nucleus at the centre and electrons surrounding it. The nucleus is composed of positively charged protons and the electrically neutral neutrons. The electrons that surround the nucleus are negatively charged. Finally, the subatomic particles were found to be constituted of the elementary particles that are the most basic building blocks of matter. It is not atom but these elementary particles that are finally uncuttable. Electron is an elementary particle, but protons and neutrons are not. They are composed of quarks.

Elementary particles can be divided into two groups: "fermions" that are building blocks of matter, and "bosons" that are carrier of forces of nature. Fermions comprise of a set of 12 particles: six "quarks" and six "leptons." Quarks participate in the strong nuclear force that holds the nucleus, and therefore are constituents of the protons and the neutrons. The six quarks are whimsically named up, down, charm, strange, top, and bottom. Leptons may be charged (like electrons, other charged leptons being muon and tau), or electrically neutral (like three kinds of neutrinos). These fermions combine together to make all matter that we see.

As we discussed shortly before, bosons are the carrier of forces. There are four kinds of forces in nature: strong nuclear force, weak nuclear force, electromagnetic force, and gravitational force. Each of these forces has one or more corresponding exchange bosons. Strong nuclear force acts at a very short range, but is the strongest of the four forces of nature, and is responsible for holding together quarks in the protons and neutrons and keeping the nucleus intact. The exchange bosons for the strong force are called gluons. Electromagnetic force is the next strongest force, can be attractive or repulsive, has an infinite range, and its exchange boson is the photon. Weak force is responsible for radioactive decay, has a very short range, and the corresponding exchange bosons are called W^+, W^- and Z^0. Gravity is the weakest of all four forces, has an infinite range, is attractive for all kinds of matter, and its exchange boson is postulated to be "graviton" which has not yet been observed.

In addition, we have "antiparticles", those that have the same mass but opposite charge and other properties. Thus there is an anti-electron (also called positron), anti-proton, anti-neutron, and so on. Neutrons and anti-neutrons are both electrically neutral, but are composed of quarks and anti-quarks respectively which have opposite charges. If a particle and its corresponding anti-particle meet, they annihilate by producing pure energy.

Energy

Energy is the capacity of a physical system to perform work, such as causing motion, or interaction between molecules. Energy may be associated with matter, or may exist independent of it. Conventionally, energy was described as kinetic energy and potential energy. Simply put, kinetic energy is the energy of a body associated with motion, and potential energy is the energy contained within the body at rest. For a body of mass m moving at a uniform velocity of v, kinetic energy (KE) can be mathematically calculated to be: $KE = \frac{1}{2} mv^2$. Similarly, if a body with mass m is placed at a height h above the ground, its potential energy (PE) with respect to the ground is given mathematically as: $PE = mgh$, where g is the acceleration induced by the gravitational force of the earth. For any object in a system, the sum of its kinetic and potential energy should always be conserved. This is best illustrated by an example of a simple pendulum that oscillates from one end to another (*Figure 4*). As the pendulum reaches one end, its height above the baseline mean is the maximum, hence the potential energy is the maximum, and thereby the kinetic energy is the least. In fact, the pendulum should stop momentarily before reversing the direction of its swing, hence the kinetic energy of the pendulum at that end becomes zero when its potential energy is the maximum. As the pendulum continues to move towards the baseline, its velocity increases, hence the kinetic energy increases. The potential energy would decrease proportionately and at the baseline when the velocity and hence kinetic energy is the maximum, the potential energy would become zero. When the pendulum continues to move towards the other end, its velocity and hence kinetic energy keeps decreasing while its potential energy keeps increasing till it reaches the other end when the kinetic energy becomes zero and potential energy is the maximum. The process repeats itself indefinitely in the form of pendulum swinging from one end to another if the friction associated with the air is somehow reduced to zero, or if the pendulum is placed in vacuum where the effect of air friction is negated.

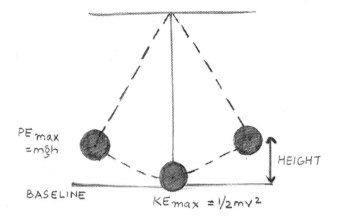

Figure 4: Swinging pendulum: An example of the conservation of energy while the kinetic energy is being converted to potential energy and vice-versa.

The classification of energy into kinetic and potential energy can be broadened to include various other forms of energy. Potential energy is the energy stored in the body at rest and not shown through motion. Examples of potential energy would include chemical, nuclear, gravitational and elastic. Kinetic energy is the energy associated with motion, and examples of kinetic energy would include mechanical, thermal, sound, electrical, and electromagnetic. The illustration of an oscillating pendulum is an example of potential energy gained by the gravitational attraction of the pendulum towards the earth and kinetic energy associated with the motion of the pendulum. We shall briefly consider other forms of potential and kinetic energy.

Chemical energy is a form of potential energy that is stored in the chemical bonds between atoms of a molecule. It is because of the interaction of the electrons of the atoms that form a molecule. When a chemical bond is formed, energy is absorbed, and when it is broken, energy is released. Chemical reactions that release net energy are called exothermic and those that require energy are called endothermic. It is worthwhile remembering that all these chemical reactions are interactions between the electrons of the outer shell of the participating atoms; the nucleus and even the electrons in the inner shell take no part in the formation or dissipation of chemical bonds.

Nuclear energy, on the other hand, is a form of potential energy that is stored within the nucleus. The arrangement of protons and neutrons within the nucleus stores this nuclear energy. Tremendous amount of energy can be released in nuclear reactions that may involve fission or fusion of individual nuclei. While in a chemical reaction, only the outer electrons rearrange themselves and the individual atoms retain their identity, in a nuclear reaction, the individual atoms lose their identity and are transformed into other atoms because of a change in the structure of their nuclei. For example, in a nuclear fission reaction involving the bombardment of uranium-235 with a neutron, an unstable isotope uranium-236 is produced which breaks down into fast moving smaller nuclei (Kr-92 and Ba-141) and three free neutron, thereby releasing a huge amount of energy. These free neutrons can then act on the other nuclei of uranium-235 and thus a self-sustaining chain reaction sets in. When uncontrolled, this chain reaction can be used as an extremely destructive nuclear weapon, the sort of "atom bombs" that were dropped on Hiroshima and Nagasaki in the Second World War, causing immense death and destruction. In contrast, a controlled nuclear reaction can be used in a nuclear reactor for useful purposes of generating power and electricity. Another way of harnessing nuclear energy is by fusion reactions in which two nuclei fuse to form a heavier nucleus, releasing a huge amount of energy in the process. For example, when two unstable isotopes of hydrogen, deuterium (H-2) and tritium (H-3) are accelerated to great speed and collide with each other, the electrostatic repulsion between the positively charged nuclei is overcome and these two nuclei (H-2 and H-3) fuse to form helium (He-4) and release a free neutron and a tremendous amount of energy in the process. Almost the entire energy produced in sun and stars is because of nuclear fusion. While fission reactions have been used in a controlled manner as well for peaceful production of electricity and other forms of energy, the attempts to control nuclear fusion reaction have not yet been successful.

Gravitational energy is a form of potential energy that is built up because of gravitational attraction of one body to another. The build-up of the potential energy of the swinging pendulum as it moves farther from the equilibrium is an example of this form of energy. Similar potential energy would be stored in planets and stars because of their gravitational attraction. So there would be a gravitational potential energy of the earth-sun system, of the sun-milky way system, and so on.

Elastic potential energy is a form of energy that is built up as a result of elastic deformation/ extension/ compression of an object. On an atomic scale, this elastic energy is due to the reversible strain placed between the bonds of the atoms, without any permanent change to the object. For example, it is built up when a spring or a rubber band is stretched and is released when we let go. Another example is when we drop a tennis ball on the ground. As it impacts the ground, its kinetic energy gets converted into an elastic potential energy due to the deformation of the ball on hitting the ground. This elastic potential energy is again converted first into kinetic energy when the ball bounces back and then into gravitational potential energy as the ball gains height. Energy conversion to sound and heat due to inelastic deformation and air resistance would result in the successive bounces less than the previous one. The elastic potential energy also finds application in sports. In archery, when the bowstring is pulled, elastic potential energy is built up. When the archer releases the bowstring, this elastic potential energy would be transferred to the arrow as kinetic energy and the arrow flies.

Sound and thermal energy are broadly, forms of kinetic energy. Sound is the result of mechanical vibration that propagates through the air or other media. When one speaks, the molecules in the air participate in a vibration that forms the sound wave. This wave travels to another person and impacts on his eardrum. A series of mechanical, chemical and electrical connections from the eardrum to the brain occur and we perceive it as hearing. All musical instruments are based on this principle of generation and propagation of sound waves. Certain patterns and frequencies of the sound waves are perceived as music, while other patterns and frequencies are perceived as noise. Our capacity to hear the sound waves is limited to certain range of frequencies, 20 to 20,000 hertz. Beyond these frequencies, we can't hear. As discussed earlier, dogs have a capability to hear the sound with frequencies ranging from 50 hertz to 46,000 hertz, while bats can hear frequencies between 3000 hertz and 120,000 hertz. Much higher frequencies of sound waves (1 – 20 mega-hertz) find application in medical imaging, in the form of ultrasound. These waves propagate through the body and are reflected by the organs within. The reflected waves are then received by the hand-held transducers and are converted into visual images that are interpreted by the radiologists.

Thermal energy is the energy associated with random motion of the atoms and molecules constituting the substance. In a gas, the kinetic energy associated with the motion of the constituent molecules including their vibrational and rotational movements. In case of liquids and solids, the potential energy of inter-atomic attraction is also involved in addition to the kinetic energy of the constituent atoms. Thermal energy is what gives heat and temperature to the object. When there is a difference in the temperature between two objects in contact, thermal energy flows in the form of heat from the hotter object to the cooler in order to achieve an equilibrium temperature. The entire principle of cooking is based on the transfer of heat.

Electrical energy is the energy associated with the moving electrons. When there is a difference in electric potential (voltage), an electric field is generated and this gives rise to the flow of electrons. In neutral conditions, the electrons are supposed to be bound to their respective atoms. However, when there is an electric field applied to a conducting element (a copper wire for instance), there is a unidirectional flow of electrons in that element. This flow of electrons is electricity, and the energy associated with it is electrical energy. This form of energy is one of the most popular forms of energy utilized in day-to-day activities. The electrical energy can be converted to light, heat, sound, mechanical energy and so on for diverse purposes. The applications include electric lights and fans, turbines and elevators, heaters and air-conditioners, refrigerators and washing machines, televisions and computers, and myriad of other uses.

The forms of energy that have been described so far apply to the association of energy with matter. When it is independent of matter, it is associated with electromagnetic waves. These waves (for instance, light) carry energy with them. The amount of energy contained in the electromagnetic wave is proportional to the frequency (f) of the wave, and is given by the equation, E = hf (h being the Planck's constant). This simplistic description is not completely true as we know now that light is composed of photons which have dual characteristics of both particle and wave. Photons can be considered to be packets of energy that travel as light that again is a form of electromagnetic wave. Each individual photon is associated with a small amount of energy, and light can thus be considered as propagating group of such photons. When light strikes an object, some photons may get reflected back to travel in another direction, while other photons may dissipate their energy in some other form. At such small scales as that of a photon, the distinction between particle and wave gets blurred and each photon behaves as both a particle and a wave.

Space

Space as we know can be described as a three dimensional space. There have been excellent descriptions of the geometry of such a three dimensional space. Matter occupies space, the shape of a matter conforms to the dimensions of the space; the movement of matter takes place within this space. When we think about the universe, the picture that comes to our minds is that with dust and pebbles, land and oceans, earth and moon, planets and sun, stars and galaxies; all contained within the huge space within the universe. So space is the home of the material constituents of the universe. The entire drama of the universe that we see and imagine takes place within this space. Even where there is no matter and no energy, there would be space. Even vacuum occupies space. There are vast regions of vacuum in the intergalactic space. Nothing can be comprehended in the absence of space.

In olden times when man looked around himself and also towards the skies, he tried to comprehend nature by ascribing shapes and trying to find correlations and patterns in the shapes that he saw. This was the birth of geometry. It started with straight lines, circle, triangles, rectangles, squares, and other shapes that could be drawn on a 2-dimensional flat surface. Very soon, this geometry was extended to include three dimensions that would be a more complete description of the three-dimensional space and its constituents. Other shapes like sphere, cylinders, cones etc were accurately described and the rules of geometry were neatly laid down. This kind of classic geometry of two and three dimensions came to be known as Euclidean geometry, named after Euclid, the ancient Greek mathematician who is also referred to as the father of geometry.

Whether space is an entity in itself or is just an abstract concept of the relation between different entities is another debate in philosophy of space and time. The presence of matter implies a certain location to it and this location can be objectively determined by the co-ordinates of three-dimensional geometry. Different bodies would occupy different locations and the intervening region between these locations is a relation that may be termed as space. The German mathematician Leibniz was the champion of such an abstract concept of space. According to this view, space exists because of matter; if there were no matter, there would have been no concept of space. On the other hand, Isaac Newton maintained that space is an absolute entity in itself, quite independent of the presence of matter within it.

When one looks at the concept of space in the context of the Big-Bang theory of the origin of universe, it is difficult to judge which of the two arguments is true. All the matter and energy of the universe was concentrated in a tiny region, and the primordial explosion of this highly dense matter-energy complex resulted in the creation of the universe. The universe expanded very rapidly and this resulted in the distribution of matter and energy throughout the space. In fact, the universe continues to expand even now. Expansion of universe must necessarily mean expansion of space. The total volume of space at present must be billions of times more than the total volume of space immediately following the Big-Bang. Thus, while space is a result of the distribution of matter within the universe that happened after the Big-Bang; it is also a necessary requirement for the existence of matter.

Time

The concept of time has intrigued philosophers and writers over centuries. The ancient Greek philosopher Plato considered time as "a moving image of eternity; and this is a register that distracts no one from the conviction that eternity is an image made with substance of time…" Plotinus, another prominent Greek philosopher and mystic, maintains that "… time is a diastasis – a spreading out – of the soul … It was necessary for the soul to create space, so that it might go forth; time was analogously born from the desire of a restless faculty of soul that wished to command itself and be of itself and acquire more than what it had."

In his celebrated book "Confessions," St Augustine muses, "What is time? … The future is not yet here, the past is no longer here, and the present does not remain. Does time, then, have a real being?" He continues… "But when it is measured, where does it come from, by what path does it pass, and whither go? Where from, if not the future? By what path, if not the present? Whither, if not into the past? It comes, then, from what is not yet real, travels through what occupies no space, and is bound for what is no longer real." In Augustine's view, time is both real and unreal; it exists and also does not exist.

Leo Tolstoy, the celebrated Russian writer, wrote in his diary that was later published as "Who, What am I": "The past is that which was, the future is that which will be, and the present is that which is not. That is why the life of man consists of nothing but the future and the past, and it is for the same reason that the happiness we want to possess is nothing but a chimera, just as the present is." Note Tolstoy's definition of present as "that which is not." When you realize it, the present has already become the past. Is time real?

No wonder, then, that of all the basic components of the universe, time is the trickiest to understand. We know of it only by its passing away. It has a direction that proceeds from what we call as past to what we call as future. Present is a slice within this continuum from past to future. We cannot measure time as it is; we only measure the passage of time. We would not know what time is if it were not for the continuous unrelenting passage of time. Time can be combined with other physical quantities like mass and distance to generate the concepts of speed, motion, energy, momentum and so on. Similar to space, time is the essential requirement of all the drama that happens in the universe.

Just like space, time too could be considered both as an entity in itself or just an abstract concept signifying relations between different events. In the Newtonian version, time is an independent absolute entity in itself; it flows uniformly of its own nature without any external influence. It could be also thought of as just an abstract concept for the convenience of recording and measuring events that take place at different intervals. In the last chapter, we have also seen the difficulties with the view of absolute time, especially with Einstein's theory of relativity.

When we look at the concept of time in the light of the Big-Bang theory of the origin of universe, we are faced with several tricky questions. Big Bang happened about 14 billion years ago and this primordial explosion could be considered as an event that led to the birth of the universe. The physicists suggest that because the universe was born at that moment, time too began then; and any question relating to the events preceding Big Bang is not permissible in the context of the existing universe. According to them, one of the important lessons of the Big Bang theory is that time had its birth at the moment of the bang. But is "time" dependent on the event of the Big-Bang? Or does it have an existence independent of such an event? Does time precede the existence of the universe? These are questions that do not have a clear answer in science.

Another peculiar characteristic about time is that it is unidirectional, that is there is an arrow of time that always points from the past to the future. In contrast, the three dimensions of space do not have any fixed direction. Of course, this unidirectional arrow of time forms the basis of the phenomenon of cause and effect, and seems too intuitive to be questioned. But, is there any specific reason why time should have this unique feature? The second law of thermodynamics is often used as a justification of this arrow of time. According to this law, the entropy of a system would never decrease. Thus, when a glass jar falls from the table on the ground and breaks into several pieces, the entropy of the system increases. If time were to flow in the reverse direction as well, it would mean that the pieces of glass jar would reassemble to form the jar and then get somehow elevated from the ground and occupy its place on the table. This process not just sounds as ridiculous, but is also not possible because it would involve a decrease in the entropy of the system. Thus time can only flow in a forward direction, from the past to the present to future.

Time and space seem to be a priori requirements of the existence of the universe. Yet, when we confront the situation at the moment of the Big Bang, it seems that they were created at that particular moment. Since the Big Bang, time has been flowing and space has been expanding. Nobody questions as to where from the time and space are being formed? It seems so natural for us to just accept the continual creation of time, and the physicists have a very good explanation for the continual expansion of space. Where does the law of conservation stand as regards space and time? When we talk about matter and energy, together they must be conserved. Why doesn't the same principle apply to space and time as well? These are some uncomfortable questions that have seldom been asked and that we shall attempt to discuss these in the next chapter.

6

Law of conservation

"Classical thermodynamics ... is the only physical theory of universal content concerning which I am convinced that within the framework of the applicability of its basic concepts, it will never be overthrown."

■ Albert Einstein

If there is one law that is intuitively the most accurate and should stand the test of time, it is the law of conservation. At the most basic level, the law of conservation of mass states that the total mass of a system should remain constant, despite the matter changing form. The law of conservation of energy (the first law of thermodynamics) states that energy can neither be created nor destroyed; it can only be changed in its form. It would also imply that the total amount of energy in the universe is always the same. With Einstein's assertion about the equivalence of matter and energy with his famous equation, $E = mc^2$, this law of conservation is extended to conservation of mass and energy. Mass, according to Einstein, is nothing but condensed energy. The process of nuclear fission releases a tremendous amount of energy locked in the form of mass. In the process, the total mass of the individual components participating in the nuclear reaction will be reduced in exact magnitude of the amount of energy released. The total magnitude of mass and energy would remain constant.

If we imagine the universe to be one big closed system, the law of conservation implies that the total amount of mass and energy of the universe will always remain constant. There is a particular magnitude of mass-energy that has been existent in the universe since the time of its birth, and this magnitude must remain constant. But what was the origin of that magnitude of mass-energy complex is a question that remains unanswered.

So much about the conservation of mass and energy. Now let us move to the other basic physical components that make the universe; that is space and time. From the Big-Bang theory of the origin of the universe, we know that the entire matter and energy of the universe was concentrated in a single point, the so-called singularity. The explosion of the primordial fireball, if one may call it so, caused the expansion of space and distribution of matter and energy throughout the space. The initial moments following the Big Bang saw a tremendous rate of expansion of space. As the space expanded further, matter and energy were distributed throughout the newly formed space. There were small differences in the density of the space, and the clumps of matter and energy in the newly formed space became the building blocks of the galaxies and stars. In fact, the universe seems to be expanding even now. In the middle of twentieth century, Edwin Hubble observed that the spectrum of radiation emitted by distant galaxies are shifted towards the red end of the light spectrum (redshift), that is there is a uniform reduction in the frequency of the radiation emitted by those distant galaxies. In terms of Doppler effect, this would imply that the distant galaxies are receding away from us at a considerable speed. Hubble also observed that the magnitude of the redshift is greater for galaxies that are at a greater distance from us, meaning thereby that the distant galaxies are moving away from us with speeds that are proportional to their distances from the earth. This is consistent with Hubble's picture of expanding universe in which each galaxy is receding away from every other galaxy because of uniformly expanding space. When extrapolated backward in time, this would lead to a moment when all the matter and energy of the universe were concentrated at one point; rather the entire universe was contained in an extremely dense point of singularity, the primordial fireball. This was a very important argument in favor of the Big Bang theory of the origin of the universe, which says that the explosion of this primordial fireball led to the creation of the universe.

As we have discussed earlier, time is perceived and measured by its flow. Time flows from past to future, through the present moment. In terms of the Big-Bang theory, time began when the universe took birth at the moment of the big bang. Since then the flow of time has been incessant and uninterrupted. The flow of time is so obvious and matter of fact for us that we don't even think about it. This sacrosanct view of uniform flow of time was challenged in the beginning of twentieth century by a young physicist, Albert Einstein. In his special theory of relativity, Einstein shattered the myth of absolute ideas of space and time. He explained that it is neither space nor time that enjoys the special status of absoluteness; rather this special status is held by speed of light. The speed of light in vacuum is constant across all frames of references and our measurements of distance, time and mass are relative. For a stationary observer, an object moving close to the speed of light would appear to be heavier (its mass would be more), shorter (length contraction) and its clock would run slower (time dilation) than when the same object is stationary or is travelling at a considerably lower speed.

Further refinement of the concepts of space and time led to the conclusion that they are actually not separate entities, but are closely knit and intertwined as "space-time." Space has three dimensions and time can be considered as the fourth dimension of the space-time.

So far, so good. The problem with all this explanation about space-time is this: why does the law of conservation not apply to space and time? This law of conservation seems to be so obvious and intuitive when it concerns mass and energy. If someone would suggest that a particular building just appeared out of blue, our scientific temperament would rebel against this idea. The total amount of mass and energy of a system would always be conserved; the forms may change. Now let us consider the whole universe as one system. Matter and energy together must necessarily be conserved, and this statement seems so intuitive. Yet, when we talk of space and time, why should the same requirement of conservation not hold true? Space has been expanding right from the big bang and continues to expand even now. Time keeps flowing from past through present to future, generating itself at every moment. Space and time keep being continuously generated out of nothing; and no doubt is ever raised on this.

There is no a priori reason to assume that the law of conservation should hold only for matter and energy and not for space and time as well. If this point is examined purely logically, we should be applying the law of conservation to the entire universe as a whole, with its constituent components of space, time, matter, energy and other unrecognized and yet unexplained entities like consciousness etc. Why should this very important principle then be limited to mass and energy, and why should it lose its application when other entities like space and time are considered? If there is one a priori principle of logic and science, this law of conservation should hold its rightful place there. When this law is now seen in its broadest perspective, the total magnitude of everything that constitutes the universe must be always conserved. All the entities, viz matter, energy, space and time must be included in this widest perspective of the law of conservation. But how do we consider the entities that have absolutely different units together? How do we add the entities as different as space, time, matter and energy? Only when we manage to perform this seemingly impossible task of combining these diverse entities, we can talk about the sum total of everything in the universe. Only then can we assert that this sum total would always be constant according to the law of conservation applied to the universe as a whole. This task of combining such different entities like space, time, matter and energy is something much weirder than comparing apples and oranges.

Or, are these entities really as different as they seem to be? Is there an underlying unity among the different components of the universe? Do they have a common origin? About a hundred years ago, it would seem ridiculous to suggest that space and time are not separate entities, that mass and energy are equivalent. With his theory of special and general relativity, Einstein showed that space and time are not really different from each other; these two are considered together as a four-dimensional space-time. One of the other consequences of the theory of relativity was establishment of the equivalence of mass and energy; the equation that demonstrates the mass-energy equivalence ($E = mc^2$) is perhaps the most famous equation of science among the lay public. Today, almost all students of science will have some idea about the basic unity of space-time, and that of matter-energy.

This idea of this basic unity needs to be expanded to all the components of the universe. My suggestion simply is that space-time on one hand and matter-energy on the other are actually not unrelated entities; they are different manifestation of the same underlying fabric of reality. Together, they must obey the law of conservation. This suggestion of unity of space, time, matter and energy may seem weird, but I am convinced that this is the only way that we can truly uphold the law of conservation in its broadest sense. I shall expand on my premise in the next chapter.

7

Interplay between the
basic components of the universe

"We can imagine that this complicated array of moving things which constitutes 'the world' is something like a great chess game being played by the gods, and we are observers of the game. We do not know what the rules of the game are; all we are allowed to do is to watch the playing. Of course, if we watch long enough, we may eventually catch on to a few of the rules. The rules of the game are what we mean by fundamental physics."

■ Richard P. Feynmann

We discussed previously about the four basic components of the universe: matter, energy, space and time. All these four components appear to be quite distinct from one another in our daily life and also in most of the scientific considerations. The interplay of these four components can be seen in all kinds of motion and forces that we come across. For example, a body in motion has some speed, and speed is distance covered per unit time. Speed or velocity shows an interaction between the measurements of space and time. When a force is applied to a body in rest or uniform motion, it gains speed, that is, it accelerates. In accordance to Newton's second law of motion, the force (F) required in such a situation is directly proportional to the mass (m) of the body and the acceleration (a) achieved, and is given by the equation, $F = ma$. Force takes into account the mass of the matter, as well as the acceleration which in turn is change of velocity per unit time. Hence force shows the interaction between mass, space and time. Similarly, the concepts of energy, work, and momentum are examples of interplay between these basic components that make the universe.

These are examples of the interactions between space, mass, energy and time without any alteration in the basic structure of these entities. These four basic entities retain their distinct properties and the concepts of velocity, momentum, force, energy, etc. are merely convenient descriptions of the various aspects of their mutual interactions in nature that do not really change the intrinsic structures of these basic entities. However, there is a problem about this simplistic understanding of space and time, matter and energy, if we bring the principle of relativity into this discussion. When we look at nature from the viewpoint of the theory of relativity, we find that there is a greater unity among these basic components of the universe and that they are not such distinct entities as they appear to be.

I must confess that I do not have a great understanding of the mathematics behind the theories of special and general relativity. I have tried to understand them in the context of my thoughts and would try to explain this in my own amateur way. There are certain conclusions drawn from Einstein's theory of special and general relativity:

1. Space and time are not fundamentally distinct entities. Rather, the three dimensions of space and time are actually united to form a four dimensional fabric of space-time.
2. The mass and space dimensions (say length) of a particular object are not constant; rather it depends

on its motion. Even the passage of time is perceived differently for the same object depending on the speed with which it moves. As has been explained by this theory, for an object travelling at speed near the speed of light, time dilates, length contracts, and mass increases. Such an object will become smaller in length, greater in mass, and would experience time to slow down (the clock will run slowly for this object).

3. Mass and energy are interchangeable. Remember the famous equation $E = mc^2$, where E represents energy, m represents mass, and c is the speed of light.

4. Nothing can travel faster than the speed of light (c). This would mean that the maximum speed permitted by nature is c. This has a tremendous implication in science and mathematics, which has perhaps not been fully understood.

One of the most stunning revelations of the theory of relativity was the alteration of the concepts of absolute space and absolute time. Till then, physics was based on the idea of an absolute space that has three dimensions, and an absolute time that flows uniformly. The absolute nature of space and time was considered to be independent of the existence of matter and energy. With the advent of special relativity, the geometry of the three-dimensional space gave way to the four-dimensions of space-time. Space is no longer three-dimensional and time is no longer a distinct entity. Instead both space and time are now considered to be intertwined and inseparable, forming a four-dimensional continuum of space-time. The repercussions of this conclusion are startling. Every event happens in a slice of space-time fabric. To different observers, two events may have a different temporal sequence. Two events that would appear to be simultaneously occurring to one observer may appear to happen one after the other to another observer in a different inertial frame of reference. The absoluteness of space and time being demolished, a new fundamental constant takes its place; this being the speed of light in vacuum (c). In fact, the whole idea of special relativity dawned upon Einstein as a result of his thought-experiments with objects travelling at speed close to that of light. The speed of light is unsurpassable; nothing can travel faster than light. This is the maximum speed permitted by nature.

Another remarkable consequence of the theory of relativity is the equivalence of mass and energy. Remember the equation, $E = mc^2$, where E is energy, m is the mass of the particle, and c is the speed of light in vacuum. This is arguably the most famous equation in physics, and perhaps in the whole of physical sciences, and even lay people with limited or no background of physics would have at least heard of this equation. Now, this equation implies that a small amount of matter has stored in it a tremendous amount of energy, because of the equivalence factor of the square of the speed of light, which is a very large factor. Considering the speed of light to be 3 hundred thousand kilometers per second, one can imagine the scale of the magnitude of energy stored in a particle with a small mass. For example, one gram of sand-grains would contain within it 90 billion kilo-joules of stored energy that is approximately equal to 21.5 million kilo-calories worth of energy. However, this energy is not easy to be released from matter. A small fraction of the mass is converted into energy in nuclear reactions and yet, these nuclear reactions are such potent sources of energy. The world has seen its application as a destructive force in the form of nuclear weapons as well as a useful way as source of power and electricity. Such conversion of mass into energy is also encountered in particle physics during the annihilation of particle and anti-particle with release of pure energy in form of photons. For example, an electron and a positron (positively-charged electron with the same mass) may collide and annihilate each other with the production of energy given by the equation $E = mc^2$. The conversion can also be in a reverse fashion wherein energy may be converted into mass. In particle accelerators, when force is applied to accelerate a particle close to speed of light, the speed of the particle cannot increase indefinitely; it is limited by the speed of light that is the maximum allowable speed. Much of the energy transferred to the particle moving at such high speed gets converted into mass, and thus the mass of the particle would increase. The particle gets heavier and heavier as it gets accelerated close to the speed of light; so much so that for it to achieve the speed of light, its mass would increase to infinity. Since it is not possible for a particle to attain an infinite mass, the speed of light is not breached.

From the preceding discussion, we see that mass and energy are two aspects of the same reality. Hence, among our fundamental entities (space, time, matter, and energy), two (matter and energy) are not really different from each other. We have also seen that space and time are not really distinct from each other; rather they are intimately connected and integrated to form a four-dimensional fabric of space-time. This ultimately reduces the four basic entities of the universe to two: space-time and matter-energy.

We have so far applied the consequences of Einstein's theory of relativity to prove the unity of matter and energy on one hand, and space and time on the other. Let us move a little further into the relationship of these two real basic ingredients of the universe: space-time and matter-energy. We would need to again stand upon the shoulders of Einstein and now have a glimpse of his concept of general relativity. Recall our discussion on the theory of general relativity previously. Matter interacts with the four-dimensional space-time in such a way as to distort the geometry of this four-dimensional fabric. This distortion of the geometry of space-time fabric is the underlying explanation of gravity, according to Einstein's theory of general relativity. More massive an object, the greater is the curvature of space-time induced by it, and thus the greater is its gravitational attraction. In fact, a sufficiently massive and compact object can alter the geometry and induce a curvature of space-time to such an extent that its gravitational attraction would be enormous; so much that even light cannot escape it. This is a rough explanation of a black hole. Light that can be considered as a bunch of photons (quanta of energy) is also influenced by gravity. The explanation of gravity being a consequence of the alteration of geometry of space-time by matter-energy is spectacularly elegant. The mathematics underlying it might be too complex to comprehend, but the idea is simple. And in this simplicity lay the genius of the great man.

We see, therefore, that the four basic components of the material universe (space, time, matter, energy) could be reduced into two (space-time and matter-energy) owing to the equivalence of matter and energy, and time being the fourth dimension of the space-time continuum. We have further seen the interaction between these two basic components; how matter-energy interacts with the space-time fabric resulting in gravitational attraction. Another leap of faith would be the demonstration of the equivalence or basic unity of space-time and matter-energy. **My own position on this is that matter and energy are nothing but condensations or knots in space-time. Alternatively, space-time could be viewed as the spreading out or unravelling of matter-energy.** Perhaps one day, some physicist or mathematician would be able to explain the nature of matter and energy as a result of localized condensation in the spacetime, or knots in the spacetime.

All kinds of earthenware, whether it is glass, pot, pitcher or flower-vase, is made of mud or clay. The vessel is just a form, the underlying reality is clay. In another context, whether it is a tree, flower, man, tiger, or any other living being, the underlying factor that is a necessary for life is DNA. Different living beings are just different forms of one underlying reality that is DNA. In yet another context, all matter in this universe is made up of atoms that are further made of protons, neutrons, and electrons. So, whether it is a star, planet, moon, mountain or a stone, they are just different forms of the same underlying reality that is a bunch of elementary particles that make the atoms. The conclusion that we are moving towards is that there is one basic substratum of the universe, the reality as one may call, which manifests itself in different forms of space, time, matter and energy; and the interactions between them causes various forces and phenomena of the universe (*Figure 5*). The enquiry into "reality" actually would then be the understanding of the parent of these basic components (space-time and matter-energy). Only then we can understand the origin and nature of the universe.

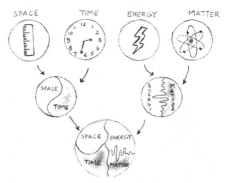

Figure 5: Common substratum of space, time, matter and energy: Space and time were wedded together as space-time by special relativity, mass and energy equivalence was established by Einstein (E = mc2). Gravity is a result of the interaction of matter-energy with space-time.

8
Theories about the origin of universe

"Some part of our being this is where we came from. We long to return. And we can. Because the cosmos is also within us. We're made of star-stuff. We are a way for the cosmos to know itself."

■ Carl Sagan

Where did it all come from? Did the universe have an origin? If so, from what did it appear, and why? Or, has it always existed? What would be fate of the universe? What is our role, as conscious beings, in this drama of the universe? These questions have stimulated man's curiosity since antiquity. Philosophers and scientists, poets and priests, teachers and students, artists and saints – every person with an ounce of curiosity has been lured by such questions, and some have attempted to uncover the nature of reality that remains ever so elusive, yet a few worthy ones are granted a fleeting glimpse of the beauty and glory of Nature. The experience of such men and women, expressed in the language of art and science, religion and philosophy, have shaped the knowledge of mankind.

There are two categories of theories attempting to explain the origin of the universe: those involving a Creator who is responsible for the creation of universe, and the other category that explains the origin without the need or intervention of the

Creator. Different religions have their own Creationist theories. Christianity believes in the Old Testament description that God created the universe including the first human beings (Adam and Eve) within six days of creation, and He rested on the seventh day. Hindu tradition is more divided on the issue of origin. There are two aspects of Hindu thought: one is mythological and the other is philosophical. The Hindu mythological tradition ascribes creation of the universe to the deity Brahma, preservation to Vishnu, and destruction/ dissolution to Shiva; Brahma, Vishnu and Shiva forming the Trinity of Gods. While the mythological tradition among the Hindus is very popular, the philosophical and scientific essence of Hinduism is embodied in the Upanishads. The opening verse of Isha Upanishad states:

"That is Whole, this is Whole. From the Whole does the Whole comes out. After the Whole is taken out from the Whole, what remains is still the Whole."

Note that there are two "Wholes" here. There may be several interpretations of this very important verse. The most accepted interpretation in Indian philosophy is that the Whole described as the cause is the unmanifest absolute pure consciousness, the Brahman. The other Whole described as the result is the manifest universe. Accordingly, the manifested universe originates from the unmanifest absolute pure consciousness or Brahman. Even after the universe has manifested out of this, absolute pure consciousness still remains. Right now, we shall not bring in the topic of "consciousness" in the discussion. We are now dealing with the material universe and will discuss about the theories that have been propounded about its origin. Sometime later in this book, when the context is right, we shall introduce the subject of consciousness and make an attempt to formulate an integrated view of our universe.

The other category of theories regarding origin includes the modern scientific theories that do not assume the role of the Creator. One of these is the "Steady-state model" according to which the universe had no beginning and has remained more-or-less the same for all time. The "Steady State model" of the universe was proposed by Hermann Bondi, Thomas Gold and Fred Hoyle, and was based on the "perfect cosmological principle" according to which the universe appears more or less the same from any point within and at every time. This would mean that we neither occupy a preferred place in the universe, nor are present in any preferred time within it. The universe would appear to be homogeneous and isotropic and essentially the same to all observers at all times. Obviously, for the universe to be homogeneous across space and across time, it could not have an origin, neither there would be an end to it. However, a static universe is impossible according to Einstein's general relativity, and observations by Edwin Hubble had shown that the universe is expanding. The expansion of the universe was incorporated in the Steady state model. In this view, new matter is continuously created from nothing as the universe expands, so that the universe retains its overall appearance over the whole range of time. The amount of matter created spontaneously was too small to be detected, about a few atoms for every cubic mile each year. Yet, the spontaneous creation of new matter is against the tenets of the law of conservation. Because of these contradictions, this theory, though appealing for its logical simplicity and elegance, has not stood the test of time. This view is no longer the dominant belief of physicists most of who believe in what is popularly called the "Big Bang theory."

The Big Bang and the origin of universe

Two significant observations suggested that the universe would have had a point of origin. One was Hubble's observation that the universe is expanding, and the other was the discovery that there is an all-pervading electromagnetic radiation coming from all directions (referred to as cosmic background microwave radiation, CMBR). Edwin Hubble demonstrated convincingly that the universe is expanding in a way that the distant galaxies are receding away from us at a speed that is proportional to their distances from us. Extrapolated backwards in time, it pointed to the fact that the entire content of the universe would have been once concentrated as an extremely dense mass-energy complex. A primordial explosion, the "Big Bang" would have led to the origin of the universe (*Figure 6*). The use of the word explosion may not go well with purists; it was indeed not an explosion but just that the entire universe was confined to a point-like singularity and has expanded since then, creating space for its expansion. Time was also created at the instant of the big-bang. However, we shall continue to use this phrase "primordial explosion" for the big-bang for a simpler understanding for the readers not conversant with the technical complexities and difficult vocabulary of the physics of big bang.

Hubble's inference of expanding universe was based on his observation of a red-shift in the spectral lines in the light emitted by distant galaxies. Recall that red color has the largest wavelength and thus smallest frequency among the light spectrum that is perceived by human eye. According to the Doppler principle, light coming from an object that is moving away from us would appear to have a lower frequency (shift to red) while that coming from an object coming towards us would appear to have a greater frequency (shift to blue/violet). The red-shift or the shift to lower frequency means that the object (distant galaxy) that is the source of the emitted light is receding from us. The farther the galaxies, greater was its red-shift. The implication of this observation was simple: the red-shift and thus the speed with which the galaxies are receding from us are greater for those that are further away from us. This significant observation was consistent with the picture of expanding universe.

Figure 6: Big-Bang: The primordial "explosion" of singularity resulted in the formation of the universe.

Another direct observation supporting the idea of the Big-Bang was the discovery of an all-pervading background of microwave radiation throughout space. This microwave radiation was discovered accidentally by Arno Penzias and Robert Wilson in 1964 while they were studying the faint microwave signal from the Milky Way. The duo detected a mysterious noise of unknown origin coming from all directions. They initially thought that these were observational errors because of pigeon droppings in their apparatus. The pigeons were trapped, and the antenna was cleaned of the droppings, but the mysterious noise continued. This was found out later to have a much more profound and interesting explanation; that of the flash origin of the universe, the Big-Bang. The CMBR is now thought of as a left-over radiation of the explosive origin of the universe from the extremely hot and dense mass-energy complex. As the universe expanded, it cooled and the background microwave radiation that is now observed is very faint, only about 2.7K or -270.3 degrees centigrade. The CMBR is a kind of snapshot of the early universe when matter started cooling down, and neutral atoms started getting formed due to combination of protons and electrons from the initially high density plasma of matter-energy, with photons released in the process. Penzias and Wilson were awarded the Nobel Prize in 1978 for this discovery, the significance of which had almost been dismissed by them as an accidental noise due to pigeon droppings.

Very quickly, there was a reasonable degree of consensus regarding the Big-Bang to be the origin of the universe. But what was the nature of this primordial explosion? As mentioned a little earlier, it would be incorrect to presume that this occurred at a point in the existing universe and the matter thus created was thrown out into space. Rather, space and time were created at the instant of the big bang. If one uses the balloon analogy, the universe has expanded just as a balloon would expand when it is blown. The balloon, when in a collapsed state, is small and contains whatever there is on it. The expansion of the balloon creates a greater and greater surface area on it. Two points marked on the surface of this balloon would recede away as it is expanding progressively

(*Figure 7*). Similarly, the universe can be thought to have originated from an extremely dense matter-energy complex confined in an extremely minute space. The Big-Bang resulted in the expansion of space and distribution of this matter-energy complex into the expanding space. Two points on the balloon can be likened to two galaxies in the universe which would recede from each other as the universe keeps on expanding.

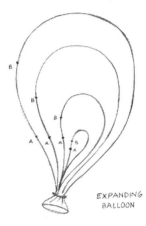

Figure 7: Expansion of space analogous to the expanding balloon. Just as any two points on a balloon move apart as it is inflated, similarly the galaxies in the universe are moving apart from one another because of the expanding space.

Although the Big-Bang theory has become the most accepted theory for the origin of universe, does it really explain the real origin? Science cannot explain the nature of the primordial fireball and what was happening within very early stages, within 10^{-43} seconds of the moment of creation (event of Big-Bang). After this time, scientific calculations can be carried out quite satisfactorily. The reason why science fails to explain what happened in this short interval after the moment of Big-Bang is that Einstein's general relativity breaks down and ceases to apply in a situation where the temperature, density and curvature of the universe are all infinite (a situation referred to as singularity).

Simply put, according to the Big-Bang theory, the universe was born out of a singularity whose nature is not amenable to scientific explanation. What happened within 10^{-43} seconds of its birth again is outside the scope of science. The later events can be fairly accurately estimated. The initial period of rapid expansion explains the bang of the Big-Bang; this expansion continues even today, about 13.7 billion years after the birth of the universe. The universe was technically born at the moment of the Big-Bang and all events before it are censured by science for which the phrase "before the big bang" has no meaning as time too was born at that moment. However, the very nature of the primordial matter-energy complex that gave rise to the Big-Bang, and the nature of the singularity that gave birth to the universe, cannot be satisfactorily explained by this theory.

So, is this theory a good theory for the origin? I would say that the Big-Bang theory explains a lot about the evolution of the universe from a state of a primordial fireball to the vast expanse of galaxies that we see today. However, it does not give a good description about the very origin of the universe. Physics ceases to apply to the very early moments of the life of the universe and thus the initial 10^{-43} seconds cannot be explained. The questions about the origin of this extremely dense matter-energy complex are not addressed. It has been said that the big-bang was a moment of singularity from which everything including matter, energy, space and time sprung up. What, then, is the nature of this singularity? Can we scientifically or mathematically explain this singularity? Modern science is almost silent on this.

In spite of this theory being so widely accepted and there being many evidence to support it, there are inherent weaknesses in it. It presumes the existence of the infinitely dense mass-energy complex from which the universe is derived. It refuses to answer or even consider the questions about the origin or nature of that infinitely dense mass-energy complex. How then is this theory different from the traditional religious belief of a Creator being responsible for the creation of the universe? The Biblical theory of creation says that God created light and darkness on the first day; sky on the second; earth, sea and plants on the third; the sun and stars on the fourth; sea creatures and birds on the fifth; animals, man and woman on the sixth; and on the seventh day He rested. The Big Bang theory says that the universe originated as a result of the explosion of the extremely dense primordial fireball that contained all the mass and energy; this resulted in the distribution of matter and energy into the expanding space. But whence was the primordial fireball created? What was the reason for the explosion? Whence does the space that is expanding since the big bang universe come from? The silence of science in these matters is a concealed admission of the limitations of this theory. In the final analysis, whether God made the universe or He made the highly dense mass-energy complex from which universe originated is one and the same because neither of these really addresses the issue of origin. Religion considers God as the first principle that is beyond comprehension; and modern science considers the primordial fireball as the first principle the nature of which cannot be explained either.

Is there any other means to approach the question about the origin of universe? Science concedes that it cannot explain satisfactorily the actual nature of the singularity, the nature and origin of the extremely dense matter-energy complex, the actual event of the Big-Bang, and what happened within 10^{-43} seconds of the Big-Bang. As we have seen, these issues form the actual crux of the problem of origin. When science fails to explain these, where do we look to? Rather than being a slave of scientific methods that have their own limitations, in such a situation, it would be better to look towards logic. According to Aristotle, "Logic means, simply, the art and method of correct thinking. It is the 'logy' or the method of every science, of every discipline and every art; and even music harbors it." Let us see if these problems can be thought logically.

Either the universe has always existed, or it had an origin. If it has always existed, we need to explain how and why it has done so without the need of having an origin. Steady-state theory assumes this; however, it fails to provide satisfactory explanation to the how and why of this hypothesis. On the other hand, if we believe in the fact that the universe actually had an origin, we must explain this origin. The Big-Bang theory, as we have seen so far, is actually an explanation of how the universe evolved from a primordial fireball stage to the present form. It fails to explain about the how and why of the origin itself.

When we start thinking about the origin, at every step of our enquiry we will be faced by the question: "where did it come from?". If we believe in God having created the universe, we need to explain the eternal nature of God or the origin of God Himself. Religions that believe in the creation of the universe by God also say that the nature of God is not comprehensible by human mind. On the contrary, if we go by the Big-Bang theory of the origin, we need to explain the nature of the singularity, the nature and origin of the extremely dense matter-energy complex which resulted in the formation of the universe. Science refuses to get into these questions by saying that the laws of physics break down and are not applicable to those early moments of the origin. Both these approaches; the creationist approach that considers the nature of the omnipresent God transcending time and space incomprehensible, and that of the Big-Bang that considers the origin and nature of the singularity incomprehensible; are similar in the sense that the question about the origin itself is not taken upfront.

Where did the universe come from? It originated from the extremely dense matter-energy complex because of the phenomenon of Big-Bang. Where did this extremely dense matter-energy complex come from? With a *regressus ad infinitum* approach, the issue of origin can only be resolved logically if we can explain the universe to come out of nothing. Let us attempt to examine this possibility.

9
Origin from nothing:
The "Split of Zero"

"Mathematics is the alphabet in which God has written the universe."

■ Galileo

We have seen the problems associated with the theories about the origin of the universe. The most accepted theory, the Big Bang theory does not really explain the origin; rather it attempts to explain the evolution of the universe assuming the existence of the infinitely dense mass-energy complex from which the universe appears to evolve. We have also seen that if one keeps on questioning "where did it come from?" the satisfactory answer can only be the one that can explain the origin from nothing. "Nothing" does not require any further enquiry as to where it (nothing) came from, provided we can define the nature of nothingness or zero.

Hence, we must first define "nothing." When we say that there is nothing in this room, there may not be any furniture or books or any visible material stuff. But there still exists air inside the confines of the room. So let us somehow take out all the air from the room and make it absolutely empty. Such a scenario is difficult to imagine in a usual life, but recall that there are vast expanses of space in the universe that are devoid of air. Such really empty intergalactic spaces do not contain any atom or molecule and can be thus thought of as void of all matter. But think deeply, does it still qualify to be called "nothing"? Even vacuum is not "nothing"; it is space that contains no matter. There are many things that happen within that empty space. Light from distant and nearby stars propagate through the space, thus carrying energy along with it. If you somehow enter such a vacuum space, you would also find that you would be subject to the gravitational forces of the neighbouring stars and galaxies. The gravitational fields transcend the matter-less or vacuum space. Time would also flow within such a vacuum space, just as it would if there was an earth-like planet within that space.

Nothingness thus is not a mere absence of matter from space. Even within vacuum, space exists, time flows, energy propagates in form of light, and gravitational fields exert their influence. Even the vacuum is a stage for myriad phenomena of nature. So what then, is nothing? In the strictest sense that we shall follow in our discussion, "nothing" implies absolutely nothing: no matter, no mass, no energy, no space, and no time. None of the fundamental entities that describe the universe, nor the entities derived from these fundamental entities, should exist in nothing. Mathematically, this "nothing" can best be represented as "zero" (0). In all the subsequent discussions in this book, we shall follow this strict definition of nothingness or zero.

As mentioned in the previous chapter, the big bang theory provided a good model of the evolution of the universe but failed to satisfy our intellectual curiosity about the real origin of the universe. I was getting increasingly uncomfortable to accept that the entire matter and energy of the universe was concentrated within a highly dense primordial fireball, the nature of which and the antecedent of which is neither known nor can it be a subject of any enquiry within the realms of science. I was even more perturbed by the continuous creation of space and time following the big bang, and the fact that there was no scientific enquiry regarding how and from where, and why should space and time be continuously generated. Why should the law of conservation apply only selectively to mass and energy; why shouldn't it also apply to space and time? The continuous generation of space and time from nothing violated the principle of conservation. If space and time could be created out of nothing, why couldn't matter and energy too spring out of nothingness? These thoughts kept disturbing me for years. I was moving towards explaining the origin of everything from nothing. And suddenly, it seemed to make sense. I called it the "Split of Zero."

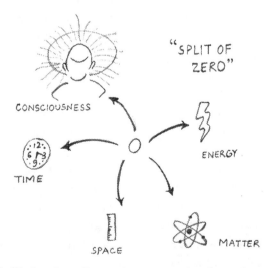

Figure 8: "Split of zero" into the entities that form the universe: space, time, matter, energy and perhaps consciousness.

Conceptually, this is a very simple theory. It is easy to understand if we replace the extremely dense matter-energy complex (of the big bang theory) with zero and the event of big bang with the split of zero (*Figure 8*). In other words, it is from nothing that everything sprang up. Mathematically,

$$0 = +x -x.$$

In this view of mine, zero or nothingness is the first principle, the singularity. Before the beginning, there was nothing; no mass, no energy, no time, no space. We can mathematically represent this nothingness as zero. The origin of universe was the result of the "split of zero." From zero, several positive and negative things sprang up. This was the beginning of time, space, matter and energy, and also of the other known and yet unknown entities and phenomena within the universe. The universe will always obey the law of conservation, that is, the sum total of all existing things and phenomena should be zero at all times from the origin through its evolution and also to its possible dissolution. It was a kind of eureka moment for me; the concept appeared very simple and elegant, and I was ecstatic. Yet, when I went deeper into the implications of the "split of zero," I came across problems with this concept.

Although this theory explains the origin of the universe from nothing, it is not without difficulties. If everything in the universe is a result of the split of zero, there must be positive and negative mass, positive and negative energy, positive and negative space, and positive and negative time. Why don't they cancel each other and become zero? If we could encounter negative space, time, energy etc., we would be annihilated. The fact that we are not, and the fact that most of the things appear to be relatively stable, attest to the possibility that probably positive and negative universes are separated by some unknown dimension. We cannot look for it because the dimension that separates these positive and negative universes is not known. We may be living in one of the bubbles of the universe and there may well be other bubbles that are negative to us and unknown. The positive and negative entities are probably compartmentalized in such a way that they do not interfere with one another. But this does not explain how and why the positive and negative entities are compartmentalized.

Some may point out the pair production and annihilation of electrons and positrons (positively charged counterpart of electron with same mass) as examples of the negative and positive entities actually interacting and cancelling each other out. But this should not be confused with spontaneous creation and annihilation of the particles in the truest sense. In particle physics, antimatter or anti-particle is a particle that has the same mass as the corresponding particle of ordinary matter, but has opposite charge and other particle properties like spin etc. The antimatter does not have negative mass. For example, positron is an anti-particle of electron: both have same mass but opposite charge (electron being negatively charged while positron has a positive charge). Another example could be a positively charged proton and a negatively charged anti-proton. When a particle and an anti-particle meet, they annihilate each other with a release of an equivalent amount of energy that is the energy equivalent of the sum of their masses (given by Einstein's equation $E = mc^2$). However, this is not annihilation in the true sense because an equivalent amount of energy is released; it is just cancelling of the negative and positive electrical charges with conversion of their masses into energy. This is not an example of matter and anti-matter actually annihilating each other in the true sense of the word.

Although they are called anti-matter, they do not have negative mass or negative energy.

There is yet another way to explain the "split of zero" without assuming the existence of positive and negative counterparts of space, time, matter, energy etc. The different basic entities may be such attributes of the universe that are inter-related in a way that their sum total is zero. So, if we consider zero or nothingness as the first principle, there was a split of zero into the entities of space, time, matter and energy. These basic entities that comprise the universe are negative and positive with respect to one another in such a way that the sum total of space, time, matter and energy is zero. They co-exist, but have acquired such different characteristics that they do not interact in a conventional sense to annihilate one another. One can borrow the concept of pluripotentiality from biological sciences to have an idea of the premise that I am trying to develop. For example, all the cells in the human body are ultimately derived from the fertilized ovum (egg). This fertilized ovum is a pluripotent cell, which means that it has a potential to develop into many specialized forms of cells, tissues and organs. After a certain amount of cell divisions, some of these cells start getting differentiated into nerve cells, muscle cells, blood cells, cells of skin and bones, and so on. These differentiated cells do not usually have the capability of reverting back to the pluripotent cells. In a similar manner, nothingness (zero) can be considered as a pluripotential entity that gives rise to space, time, matter/ energy, etc that have now become differentiated entities. We shall come to the issues of pluripotentiality and differentiation, in the context of the unity of the basic entities of the universe, later. We have discussed previously how these basic entities comprising the universe interact with one another. Space and time have been unified by theory of relativity; the equivalence of mass and energy was demonstrated by Einstein's famous equation, $E = mc^2$. General relativity shows us how the matter-energy interacts with space-time to account for gravity. We inferred that there is a common substratum of reality; a common parent of these basic entities. It is now just a matter of connecting the dots to reveal this common parent: it must be nothingness (or zero) from which space, time, matter and energy find their origin.

There are a few other difficulties with the theory of "split of zero" that we need to explain. Firstly, if zero can split, other things should be able to split too. Mathematically,

If $0 = +x -x$,
Then, $2 = +4 -2$,
        $8 = +4 +4$,   and so on.

However, we do not see any object splitting spontaneously in front of our eyes. Generally, things are stable in spite of their internal flux. For example, a table in my room may have dynamic motion of its molecules and atoms, its protons, neutrons and electrons. However, this table does not suddenly split into fragments spontaneously. A pencil or a book does not spontaneously disintegrate into its parts. What is so special about zero that it can split, while the larger finite entities in our experience appear to be more or less stable? The theory of "split of zero" fails to explain this contradiction.

In our discussion about the concept of "split of zero," we have considered zero or nothingness as the first principle behind the creation of the universe. But, why did it split? Why should zero not remain zero; why should it split to form the universe? This is another issue that needs to be resolved if we have to take the concept of "split of zero" further. What was the reason for nothingness or zero to divide and create the basic entities that form the universe? In other words, why is there something rather than nothing? In the subsequent chapters, we shall attempt to find answers to these intriguing questions. We shall start with a further expansion on the concept of zero in the next chapter.

10
Zero: its history in philosophy and mathematics

"Mathematics, rightly viewed, possesses not only truth but supreme beauty..."

■ Bertrand Russel

In the mathematics that we learn at school, the usage of zero as a number is so common that it is almost impossible for us to believe that there was once no concept of zero in mathematics. The Greek system of mathematics, championed by Pythagoras, Plato and Aristotle, had no place for zero. The geometry of triangles, squares and circles were described quite satisfactorily without resorting to zero. Yet, there was something amiss. And this was what the Greeks and western scholars feared. They were extremely intelligent people; yet they abhorred the very idea of zero. The entire Greek philosophy was based on the axiom that there is no void. So vehement was the opposition to void in the Greek philosophy and science that anything that was seen as associated with void was shunned; and the idea of void and that of zero are not just associated, they are inseparable.

Obviously, when we say that there is no apple in a basket, or there is no person in a room, there is a reference to the idea of "not being" in the basket or in the room. This idea of "not being" can be referred to zero. Though such a concept of the number is quite limited in its scope and understanding, it gives us some idea of what zero means. This idea of "not being" did not require a symbol in the numerical system that developed as a tool for counting and exchange of goods; the number system began with the number one in almost all the civilizations, from the Egyptians to the Babylonians and from the Romans to the Greeks. But the number system that started with one was faced with problem of representing the numbers like ten (10), one hundred (100), one hundred nine (109) etc. As we can see, the use of zero here is not with the meaning of nothingness or void, but it is just used as a place value for the simplicity of representing such numbers. The Greeks and Romans used letters as the symbols for the numbers; we are quite familiar with the Roman numerals: 1 as I, 2 as II, 5 as V, 10 as X, fifty as L, 100 as C, and so on. The Babylonians first used a symbol of two slanted wedges to represent what we now understand as zero in the place value of numbers. By itself, zero did not mean anything to them; it was just a digit to ease the representation of bigger numbers.

In contrast to the Greeks and Romans, the ancient Indians did not fear the idea of void; rather they embraced it. Shunya or zero (nothingness) became one of the central themes of the ancient Indian philosophy. Zero and infinity, the two most mysterious entities in mathematics, find repeated mentions in several systems of Indian philosophy dating several thousand years back. The Vedas are the oldest compilation of human thoughts and philosophy that originated in India. Although the western view is that the Vedas were compiled around 1700 to 1000 BC, it must be remembered that the contents of Vedas were transmitted through oral chanting and memory across generations much earlier than they were actually written down. For this reason, Vedas are also called "Shruti" meaning one that is transmitted by the tradition of listening. According to some of the independent Indian researchers, the Vedas date to as long back as around /UUU BC. Whatever the chronology may be, it is quite notable that shunya (zero) and ananta (infinity) were both revered by the Indian scholars right from the Vedic period. Cosmology and mathematics, mysticism and philosophy, architecture and engineering – all were included in a single cohesive body of knowledge. Zero and infinity were as much mathematical entities as they were metaphysical and philosophical. We discussed in the previous chapter on the theories of origin of the universe about the Indian philosophical view on this. The opening verse of Isha Upanishad states: "That is Whole, this is Whole. From the Whole does the Whole comes out. After the Whole is taken out from the Whole, what remains is still the Whole." The Sanskrit word used is "Purna" that normally means "whole" or "complete" in English language. It might also mean "infinity." Replace "whole" by "infinity" and we have the property of the infinity being described in this wonderful verse. One can take or subtract any number from infinity and the number still remains infinity; even if infinity is subtracted from infinity, the result is still infinity. Look at the number system. As a whole, it is infinite. Now look at a part of this number system, let's say from the number 1 to 2. There is infinite number of rational numbers between these two numbers. The range may appear small; it is just a small part of the number line; yet it contains an infinite number of rational numbers within it. One may go to smaller ranges, let's say between the fractions ¼ and ½. Even this smaller range has infinite number of rational numbers within it. So, if we take out one infinity from another, the result that remains is infinity itself. We can have countless other examples. The set of natural numbers is infinite; the set of even numbers is infinite too. So is the set of square numbers. If we take out the set of even numbers from the set of natural numbers, we are left with a set of odd numbers that again is an infinite set. Similarly, we can remove the set of squares from the set of natural numbers; the resultant set will again be an infinite set.

Unlike the western scholars who feared zero because of the rejection of void by Greek philosophy, Indian philosophers embraced zero as they embraced infinity. In fact, shunya (zero) or void became a full-fledged branch of philosophy. These mystics rejected the materialistic view of the universe and considered it as one great illusion. The only truth to them was shunya or void. Existence was an illusion; non-existence was real. It is out of the scope of this book to try and elaborate on the various philosophical ideas that the ancient Indians had about zero, but the fact is that they accepted zero as much as, or sometimes more than any other number or entity. The philosophical acceptance of shunya or zero preceded its mathematical application in India by several centuries, if not millennia. Aryabhatta, an Indian mathematician and astronomer who lived in the 5th and 6th century after Christ used zero as a place value in the numbers; like 101, 1012, 1409, etc. Although he did not use a symbol for zero, the concept of zero is clearly evident in his place value system for large numbers. Aryabhatta made several other notable contributions in mathematics including approximation of π (pi), summation of series of squares and cubes, area of triangle and also in astronomy such as explanation and prediction of eclipses etc. However, it was Brahmagupta who first formalized the use of zero in mathematical operations. He was also the pioneer in the concept of negative numbers. Brahmagupta established the basic rules governing the use of zero; these rules were:

1. The sum of a positive number and zero is a positive number

2. The sum of a negative number and zero is a negative number.
3. The sum of zero and zero is zero
4. The sum of a positive number and a negative number whose absolute values are equal is zero
5. Zero divided by a positive or negative number is zero
6. Zero divided by zero is zero

Clearly the last rule of Brahmagupta does not stand the test of modern mathematics; zero divided by zero is undefined. Nevertheless, Brahmagupta must be credited to have finally brought zero from the realms of abstract philosophy and metaphysics to the scientific field of mathematics. Division by zero was the only error among the rules formulated by Brahmagupta; and this was finally resolved by the discovery of calculus by Newton and Leibniz. Calculus dealt with infinitesimal quantities and their approximations to zero; these infinitesimals are yet other mysterious entities in mathematics.

When the Vedic seers describe the essence of the universe as subtler than the subtlest and smaller than the smallest, parallels can be drawn with the idea of infinitesimal quantities that get smaller and smaller without reaching zero. This idea of infinitesimal has been the point of raging debate and controversy for millennia. The concept of infinitesimal has troubled mathematicians and philosophers from ancient times. It led to Zeno postulating through his paradoxes that all motion is an illusion. It also led to the development of calculus and all the bad blood that it brought between Newton and Leibniz, the two men who independently discovered the discipline of calculus. We shall be elaborating about the problems of infinitesimals, the paradoxes of Zeno (an ancient Greek philosopher before Socrates) and the other similar paradoxes, and my point of view on these in the next chapter.

11

The problem of overtaking vehicles and Zeno's paradox: a critique of continuum

"The single biggest problem in communication is the illusion that it has taken place."

■ George Bernard Shaw

We all have seen a vehicle overtaking another vehicle because of its greater velocity. Let us have a closer look at this observation. Let us suppose that a car A is 20 kilometers behind a car B at a particular moment. Both cars are traveling in the same direction, each car is moving with a uniform velocity (assume A moving at 60 Km per hour and B at 40 Km per hour). Simple algebra would predict that the car A would overtake the car B in one hour.

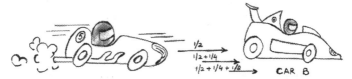

Figure 9: Problem of overtaking vehicles: The faster vehicle cannot overtake a slower vehicle that has been given a head-start unless we abandon the idea of continuum in space and time. Space and time must be discrete, not continuous.

Now, let us look at the same example in a different way (*Figure 9*). A time comes when the car A is just behind the car B. By the time the car A has moved to the position where the car B was, B has already moved a little ahead. Again by the time A reaches the position of B, the car B has again moved a little further. We can extend this indefinitely, but the car B will be always ahead and the car A will never be able to overtake it. The car A has to catch up with the car B before it would be able to overtake it. However, the distance between the two cars is getting smaller and smaller but the car A is not able to overtake the car B. There is another problem here. The time required for A to overtake B (1 hour that we calculated using simple algebra) also does not arrive. The time interval can come tantalizingly close to that 1 hour, but cannot reach there. We all know that it is contradictory to common observation and must be absurd. The car A does, of course, overtake the car B. The time interval does reach and cross one hour. But what is the flaw with this reasoning?

This problem is not a new one; it has troubled philosophers since ancient times. Zeno, an ancient Greek philosopher of pre-Socrates era, came up with a logical puzzle similar to the riddle of overtaking vehicles. In his most famous paradox of "Achilles and tortoise," the faster Achilles could never catch up with the much slower tortoise if the tortoise was given a head start (*Figure 10*). In another paradox of the "flying arrow," the arrow that is flying towards a target can actually never reach the target as it has to cover half the distance, followed by half of the remaining half, then half of the remaining quarter, and so on. It can get closer and closer to the target but there would still be some distance remaining between the arrow and the target (*Figure 11*). Zeno would thus conclude that motion is nothing but an illusion. His arguments against motion have baffled, confused, and challenged the philosophers, mathematicians and scientists for more than 2500 years. Although such a notion is clearly contrary to everyday experience, it is not easy to find flaw with such reasoning.

Figure 10:Zeno's paradox: Achilles and the tortoise

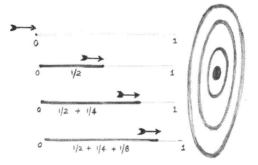

Figure 11: Zeno's paradox of the moving arrow:
He believed that all motion is an illusion.

So, what is wrong with Zeno's line of thinking? Or, as Zeno pointed out, is motion actually unreal and just an illusion? If we look into this paradox critically, it is essentially the problem of the infinitesimal. The distance keeps on decreasing by half of the previous distance; the process goes on and on. Strictly speaking, the distance can become less than any fraction that one can think of but it still must be greater than zero. This is the problem of infinitesimal: its value can be less than any number but should be more than zero; it keeps approaching zero without ever reaching there.

Mathematicians and physicists will answer the apparent paradox by using the tool of calculus. The answer provided would be that the integral of the time required for the rear vehicle to catch up with the front vehicle is finite. Similarly the integral of the distance at which the rear vehicle catches up with the front vehicle is finite too. Thus the car A would catch up and eventually overtake the car B in finite time and finite distance. The explanation may sound perfect until we actually look at the basic foundation on which the tool of calculus was built. About two thousand years after Zeno put forward the baffling paradoxes, the solution to the problem of infinitesimal was provided independently by Newton and Leibniz; it is another matter that the scientific community was and still remains divided over the relative contributions made by Newton and Leibniz. This was the birth of calculus. The concept of calculus was based on the rate of change at the ranges of infinitesimally small distance over infinitesimally small period of time. This "infinitesimally small" quantity approaches zero. Both Newton's and Leibniz's calculus are based on a very shaky foundation of dividing zero by zero, something that is mathematically unsound. At some point of calculation, the infinitesimals (or rather the squares of infinitesimals) were ignored, and correct answers were obtained. Making the infinitesimals disappear at the right time yielded correct answers, and hence calculus was embraced by mathematicians and physicists as a very important tool.

In the strictest sense, however, the infinitesimals still remain a finite quantity that is greater than zero. Approximating them to zero and then ignoring them at the right instant is the foundation of calculus, and gives answers to the paradoxes described above. For practical purposes, we have to resort to such approximations. If we stick to dividing the distances and time by half indefinitely and refuse to ignore the infinitesimals, calculus would fail to provide us with the explanation of how the faster vehicle would overtake the slower one, or in the case of Zeno's paradox, how Achilles could overtake the tortoise, or the arrow strike the wall. So, how do we explain the contradiction between the strict approach to motion as thought by Zeno and the kind of motion that we see in the actual world?

My own view is that the contradiction exists because of our assumption of continuum. The concept of continuum runs into difficulties due to the diminishing values of infinitesimals that come closer and closer to zero, but cannot quite reach zero. Unless an approximation is made, as in calculus where the vanishingly small infinitesimal quantities are actually made to vanish and approximated to zero, this riddle cannot find a satisfactory explanation. If we disregard the concept of continuum in space (distance) and time, this paradox no longer remains a paradox; it can be quite easily explained.

We take for granted the concepts of continuous space and continuing flow of time. This assumption, in my view, is incorrect. Let us take distance between two given points in space to illustrate this. As the two points A and B are brought closer to each other, the distance between them would keep on decreasing. The problem is to bring them closer and closer without the points getting overlapped. We need to make the distance between A and B as close to zero as possible, without reaching zero. We are now caught in the same trap of Zeno's paradox. The only way out now is to get out of this business of continuum and move to the concept of discrete distance. This would mean that as the two points A and B come closer and closer and the distance between them becomes smaller and smaller, a critical distance would be reached when the two points cannot come any closer; any attempt to bring them further closer would make them overlap. What does this mean? This would simply mean that there is a minimum permissible dimension of a point, and nature does not permit any dimension of distance below this magnitude.

Therefore, the distance between the two cars can become smaller and smaller, but this has to reach a limit below which the distance cannot be further subdivided. This is where the infinitesimal vanishes, as was assumed in calculus as approximation. In other words, there must be a minimum length permissible by nature (minimum quantum of length). Any distance smaller than this minimum quantum of length must simply not be possible. Similarly, there must be a minimum time interval allowed by nature (minimum quantum of time), smaller than which no time duration is permissible. The continuous uniform velocity is again a flaw; all motion is in jerks. The jerks are too small to be perceived by human eyes or by any apparatus designed to measure speed; nevertheless even an apparently uniform motion is jerky in nature, at the very very small scales of distance and time duration. The distances jump by quanta and so does time. Therefore, at a critical time when the distance between the two vehicles is small enough, the greater velocity of the car A would allow it to jump more quanta of distance than the car B at the next moment (quantum jump of time) and hence overtake it.

The problem is not just for the overtaking vehicles. In fact, we can extend the argument to show that no motion would be possible if we were to stick to the idea of continuity of distance and time. The same problem would arise if we consider the movement of a ball that is thrown by a child towards a wall, or as Zeno imagined the movement of an arrow towards a target on the wall. By our equation of mechanics, we can accurately chart out the motion of the ball till it hits the wall. However, we must think in an analogous fashion to how we thought in our example of the overtaking vehicles. The ball has to cover half the distance between the child and the wall, then half of the other half, then half of the remaining half, and so on. Even when the distance between the ball and the wall may be just 1 mm, this distance can be divided indefinitely into halves. Between the ball and the wall, there will always be infinite number of points to be traversed. But, it is our day-to-day observation that the ball hits the wall.

With the assumption of continuity, the flow of time too cannot be explained. Let us consider the passage of certain duration of time, say one day. Now a day has 24 hours and this duration can again be divided into halves indefinitely. Even when the duration left for the day is 1 second, this duration can again be divided indefinitely into halves. So the time has to elapse by 0.5 seconds, then 0.25 seconds, then by 0.125 seconds, and so on. It would go on endlessly. However, our observations clearly show that time does elapse, a day is indeed completed.

The flaw with our argument in both these illustrations of dividing the distance and time duration into halves indefinitely is, again, the assumption of continuity. It is because of the discrete (rather than continuous) nature of space that the distance between the ball and the wall cannot be indefinitely divided, and thus the ball will eventually hit the wall. It is again because of the discrete nature of time that the duration of one day cannot be indefinitely divided, and a day consisting of 24 hours will eventually pass. This motion of the ball and passage of the duration of a day can be explained very easily if we abandon the concept of continuity and accept that there must be a minimum distance and a minimum time duration permissible by nature.

12
Limit of smallness

"You never change things by fighting the existing reality. To change something, build a new model that makes the existing model obsolete."

■ Buckminster Fuller

How small can the smallest be? We have discussed that the basic ingredients that comprise the universe are discrete rather than continuous. Nature has put a restriction on the minimum magnitude of space (length) and time; this concept can be extended to matter (mass) and energy as well. Can we have a glimpse of how small these minimum allowable magnitudes are? Let us see what physics tells us.

Using three fundamental constants that have important places in relativity and quantum mechanics, a new scale of measurement was developed: the Planck scale. These three constants are Newton's gravitational constant (G), speed of light in vacuum (c), and Planck's constant (h). The constants G and c are of great importance in special and general relativity, and the constant h is of central importance in quantum mechanics. In the Planck scale, the minimum measurable length is called Planck length (l_p); and the minimum measurable time duration is called Planck time (t_p). Distance and time become meaningless when magnitudes less than Planck length and Planck time are considered.

Physicists have described complicated equations for the calculation of Planck length and Planck time; we would not go into the details of such equations and proofs as their complete understanding is beyond the scope of lay persons including me. However, it would be important to know the magnitude or value of these smallest measurements of length and time duration. Planck's length is, in principle, the minimum measurable length, and no refinement in instrumentations can allow measurement of a shorter length. Similarly, Planck time is the shortest measurable time duration, and no refinement in clock-work or other instrumentation can allow measurement of a shorter duration of time. I would extend the argument further. Planck length and Planck time do not reflect limitations of measurement of distance and time at extremely small scales; rather they should indeed be the minimum values of distance and time that Nature permits. In other words, there cannot be any length smaller than the Planck length, and there cannot be any time duration that is smaller than the Planck time.

Planck length, lp has been estimated to be approximately 1.616 x 10^{-35} meters. This is an exceedingly small length, about 10^{20} times smaller than the diameter of a proton. The smallness of this magnitude of length can be imagined by the following illustration. If we consider the contrasting sizes of a 0.1mm size dot (approximately the smallest size that human eye can see without any aid) and the entire observable universe, the Planck's length is similar in comparison to the 0.1mm dot. In other words, the ratio between the sizes of the 0.1 mm dot and the observable universe is the same as the ratio between the Planck length and the 0.1 mm dot.

Planck time (tp) is the time required for light to travel a distance of 1 Planck length in vacuum, and has been estimated to be roughly 5.39 x 10^{-44} seconds. According to special relativity, nothing can travel faster than the speed of light in vacuum; thus Planck time is logically the smallest duration of time allowed by nature. At distances less than Planck length and duration less than Planck time, distance and time would lose their ordinary meaning. Planck length thus determines the least dimension of a point in space; and Planck time perhaps is the measure of the slice of time duration that determines the present moment, and distinguishes it from past and future. It would perhaps be worthwhile to recall our discussion about the limitations of the Big-Bang theory in Chapter 8. The inability of science to explain the very initial period (the first 10^{-43} seconds) of the Big Bang may be directly related to the minimum allowable dimensions of space and time. We can see that the minimum allowable duration of time (Planck time) is somewhat similar to the duration after the big bang that is unexplainable by physics. The concepts of space and time have no meaning unless they reach their respective minimum allowable values. Space and time were created at the instant of the primordial explosion and were not different from each other at that instant. They got differentiated into seemingly different entities with further evolution of the universe.

There are values estimated for mass and energy in the Planck scale as well; the Planck mass (mp) and Planck energy (Ep). However the conceptual foundation underlying Planck mass and Planck energy are different from that of Planck length and Planck time. While Planck length and time describe the minimum magnitude of distance and time allowed by nature, Planck mass and energy are concerned with the maximum mass or energy that can be localized within a point particle (an idealized particle without spatial dimension). Of course, we now know that such a point particle must occupy some space and its lower limit should be defined by Planck length. Perhaps, it would not be erroneous to assume the point particle to be occupying a space with a diameter equal to Planck length. Now, Planck mass (mp) is the maximum allowable mass within a point particle and has been estimated to be about 2.18 x 10^{-8} Kg. If a point particle occupies a greater mass than this Planck mass, its density becomes so large that it would turn into a black hole. Planck energy is simply the energy associated with a point particle with a rest mass same as Planck mass; this can be calculated by the famous equation, $E = mc^2$. Planck energy (Ep) has been estimated to be 1.96 x 10^9 Joules.

We see now that Planck mass and Planck energy do not represent very small magnitudes. Planck mass is about 22 micrograms, equivalent to the weight of a flea's egg. Planck energy would be somewhat close to the chemical energy stored within a gasoline tank of an automobile. These values correspond to the maximum allowable mass and energy within a point particle without that particle getting converted to a black hole. But our concern was to find the minimum values of mass and energy allowable by nature. While Planck scale has helped us with minimum length and minimum time, it is of little value in our search for minimum mass and minimum energy.

Let us see if we can explore other ways of determining the magnitude of minimum mass and minimum energy allowable by nature. I urge you not to be intimidated by the equations and the calculations that I would be using for the purpose of illustration. It is the principle that is most important, equations and calculations are secondary. Max Planck showed that energy can be emitted or absorbed in discrete packets or quanta; and gave the equation describing the relationship between energy per quantum and frequency of electromagnetic wave carrying that energy.

$E = hf$;   where E = energy, f = frequency, and h = Planck's constant

Also, frequency of a wave is inversely proportional to its wavelength, and this is illustrated with the equation,

$f = c/\lambda$;   where f = frequency, c = speed of light, and λ = wavelength

Therefore, $E = hc/\lambda$

Clearly, the minimum value of energy allowed by nature can be thought of as the one quantum of energy associated with the electromagnetic wave with minimum possible frequency (or the largest possible wavelength). Recall that visible light is a small part of the electromagnetic spectrum. Human eyes can perceive electromagnetic waves with wavelength ranging from 380 nm to 760 nm, and a frequency range between 400 and 790 terahertz; this is the spectrum of the visible light. Remember the components of white light when it passes through a prism; these are the same colours that are seen in a rainbow: VIBGYOR (violet, indigo, blue, green, yellow, orange, and red). Among the visible light spectrum, red is associated with the largest wavelength and lowest frequency, and thus least energy per photon; while violet has the smallest wavelength, highest frequency, and thus maximum energy per photon. However, the electromagnetic spectrum is not limited to visible light; it has a far much wider range. The spectrum of electromagnetic waves ranges from longer than long wavelength of radio waves to shorter than short wavelength of gamma rays, thereby covering wavelengths from thousands of kilometers down to a fraction of the size of an atom. The limits of long and short wavelengths of the electromagnetic waves have not been accurately defined. It is thought that the limit of long wavelengths is the size of universe itself, while that of the short wavelength is perhaps the Planck length.

Using this information, we may attempt to find the minimum value of energy allowed by nature. This should logically be the energy associated with a single photon of the electromagnetic wave with the least possible frequency (and hence longest possible wavelength). If we consider the longest possible wavelength as the size of the universe and then calculate the energy associated with such a photon, this should give us the value of the least energy allowed by nature. We run into difficulties when we consider the expanding nature of the universe and thus its ever-changing size. The physicist and cosmologists are unsure whether the universe will keep on expanding forever, or if there is a limit to its expansion. For the sake of our calculation, we shall take the present estimated value of the size of the universe, which is about 91 billion light years in diameter. This figure corresponds to the distance travelled by light in 91 billion years. Each year has 365 days, each day has 24 hours, and each hour has 60 minutes, and each minute has 60 seconds. Thus 91 billion years would be equivalent to 91 x 365 x 24 x 60 x 60 billion seconds, which is 2,869,776,000 billion seconds or approximately 2.87×10^{18} seconds. Thus the size of the universe is about

2.87c x 10^{18} meters (where c = speed of light in vacuum). I accept that the calculation of the minimum allowable energy might be erroneous; the maximum possible wavelength of the electromagnetic spectrum may be less or more than what we have assumed. But the more important issue is the concept that there should be some minimum value of energy, E (min) that is allowed by nature, based on the maximum wavelength of the electromagnetic spectrum.

E = hf (Planck's equation) and f = c/ λ;

where E = energy associated with a photon, h = Planck's constant, f = frequency, λ = wavelength, and c = speed of light in vacuum

Thus, E (min) = hc/ λmax

Considering λmax as the size of the universe (91 billion light years or 2.87c x 10^{18} meters)

E (min) = hc/ 2.87c x 10^{18} = h/ 2.87 x 10^{18} (cancelling c from both numerator and denominator)

We know the value of Planck's constant (h) to be approximately 6.63 x 10^{-34} Joule-second

Thus, E (min) = 6.63 x 10^{-34}/ 2.87 x 10^{18} = 2.31 x 10^{-52} Joules

So, the minimum energy carried by a photon comes out to be 2.31 x 10^{-52} Joules; and as we had discussed earlier, this should be the minimum value of energy allowed by nature. Based on this minimum value of energy, we can calculate the minimum mass, m (min), allowed by nature by Einstein's mass-energy equivalence formula, $E = mc^2$

$$E \text{ (min)} = m \text{ (min)} \times c^2$$

This implies; $m \text{ (min)} = E \text{ (min)} / c^2$

Replacing the known value of c (3×10^8 m/s) and the calculated value of E (min), we have:

$$m \text{ (min)} = 2.31 \times 10^{-52} / (3 \times 10^8)^2 = 7.7 \times 10^{-69} \text{ Kg}$$

This then should be the minimum allowable mass; 7.7×10^{-69} Kg. In comparison, the mass of an electron is about 9.11×10^{-31} Kg. Photons are conventionally considered to be massless (with rest mass of zero), yet this perhaps would be the mass of the lightest photon.

Thus, we have seen that the nature allows its basic constituents to be discrete rather than continuous, and that each of these entities must have a certain minimum magnitude, however small it may be. Below such magnitude, such entities lose their individual meaning. For the sake of simplicity of illustration, let us introduce different units of measurement at this scale (unimeter for length, unisecond for time, unigram for mass, and unijoules for energy), so that the minimum value of each of the basic entities of the universe is unity (1). In this system,

Minimum permissible magnitude of length = 1 Planck length = 1.616×10^{-35} meters = 1 unimeter

Minimum permissible magnitude of time = 1 Planck time = 5.39×10^{-44} seconds = 1 unisecond

Minimum permissible magnitude of mass = 7.7 x 10^{-69} Kg = 1 unigram

Minimum permissible magnitude of energy = 2.31 x 10^{-52} Joules = 1 unijoule

As we have discussed, in principle, there is no length smaller than 1 unimeter, no time duration smaller than 1 unisecond, no mass smaller than 1 unigram, and no value of energy smaller than 1 unijoule. These are extremely small magnitudes to be relevant in everyday life, but this principle is of utmost importance to understand my view of reality.

13

Granularity of nature and mathematics of continuity

"As far as the laws of mathematics refer to reality, they are not certain; and as far as they are certain, they do not refer to reality."

■ Albert Einstein

In his short book on Relativity, Einstein explains: *"The surface of a marble table is spread out in front of me. I can get from any one point on this table to any other point by passing continuously from one point to a 'neighbouring' one, and repeating this process a large number of times, or, in other words, by going from point to point without executing 'jumps'. I am sure that the reader will appreciate with sufficient clearness what I mean here by 'neighbouring' and 'jumps' (if he is not too pedantic). We express this property of the surface by describing the latter as continuum."* This is the continuum in the context of space. We can similarly also describe continuum in the context of time. By combining the three dimensions of space and the dimension of time, Einstein gave us the idea of space-time continuum.

There are excellent mathematical and geometric expositions of the idea of continuum. In Euclid's geometry, the continuum of a two-dimensional surface is very well explained. Any point on a two-dimensional surface can be expressed in terms of two coordinates (x, y); the so-called Cartesian coordinates. However, in our day-to -day experience, we see a three-dimensional space with three-dimensional objects. When one is faced with multiple dimensions, Euclidean system and Cartesian coordinates must be modified. Gauss developed a system of coordinates for multiple dimensions wherein any point can be expressed in terms of a number of coordinates. For example, a point in three-dimensional space may be expressed as three coordinates (x, y, z); these representing the three coordinates of space (length, breadth and height). The four-dimensional continuum of space-time can be expressed as (x, y, z, t); x, y, z representing the three coordinates of space and t representing the coordinate of time (*Figure 12*). The Gaussian system can be extended to multiple dimensions. In this system, every point of a continuum is assigned as many numbers (Gaussian coordinates) as the continuum has dimensions.

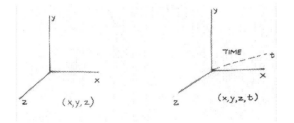

Figure 12: Co-ordinates for space-time: three co-ordinates for space (x, y, z) and the fourth for time (t) can describe the four-dimensional space-time.

In all this discussion, the theme of continuum remains unchanged; whether we are dealing with a two-dimensional surface, three-dimensional space, four-dimensional space-time, or a potentially multi-dimensional universe. How close can the two neighboring points in a continuum be? The problem of infinitesimally small distance between two neighboring points has been dealt mathematically by calculus. As we have seen earlier (Chapter 11), the assumptions made in the basic foundation of calculus disregard and ignore the infinitesimal quantities at a convenient step of calculation. When we study this problem (of how close two neighboring points on a continuum be) logically, we are stuck by the fact that for any two points, however close they may be, there would be infinite number of points between them. So how is it possible for any object to move from one point to another? Unless there is a minimum magnitude of distance, below which two neighboring points cannot come closer, no motion would be possible between any two points. One has to resort to "jumps" between two neighboring points, something that Einstein had tried to avoid while describing the continuum. Recall our discussion about the overtaking vehicles and Zeno's paradoxes. We had also extended the argument to show the existence of minimum permissible magnitude of space, time, mass, energy, and so on.

Einstein had hoped that the readers would not be too pedantic. He himself, however, came up with his brilliant theory of relativity because he was too pedantic while analyzing the Newtonian mechanics. While trying to make a theory that can explain the working of nature in a generalized fashion, one needs to be attentive to and not ignore the minute details. My position here is that the so-called continuum is not possible. It is a very useful concept for mathematical idealization, but continuum cannot exist in any dimension. We have discussed this earlier; and with the risk of being repetitive, we shall elaborate this a little further.

Let us analyze our hypothesis in a two-dimensional plane. In this plane, Euclid's geometry should hold and Cartesian coordinates can determine the position of a particular point. We shall ask this question again: "How close can two points get to each other?" If the two points on a two-dimensional plane are represented by Cartesian coordinates as (x, y) and (x1, y1), there should be an infinite number of points between these two points. The closer these two points are brought to each other, the distance from x to x1, and y to y1 keep diminishing, and yet there would be an infinite numbers of points between these two points. If there is a moving object placed at one point in this two-dimensional plane, let us analyze its movement between two points. Similar to the logic used in our discussion on overtaking vehicles, we can argue that the moving object would never be able to reach its destination; it may get closer and closer and yet, there would be infinite number of points for it to traverse. We can extend the same argument for the movement of an object in a three dimensional space or four-dimensional space-time, and reach the same conclusion that the object can never reach its destination. This argument is quite obviously contrary to our observation, and must be flawed.

What is the flaw in the argument? Again, the flaw is the presumption of continuum. As discussed in previous chapters, the distance between two points cannot be reduced below a minimum distance. One can invoke calculus to solve this problem but, as we have seen earlier, calculus is based on the logic of the vanishingly small (infinitesimal) quantities being ignored at a convenient step of calculation. If we are to strictly believe in the concept of continuum, these infinitesimal quantities can no longer be ignored. Use of calculus and the idea of continuum are mutually contradictory.

Calculus is a useful tool in mathematics and has a great deal of practical applications. The approximations that are inherent in calculus were made by Newton and Leibniz as a matter of convenience. These are approximations if one believes in the idea of continuum. But as we have discussed, the nature is not continuous; it is rather discrete or granular. In such a situation, there is a minimum value for each entity and thus, infinitesimals are not possible. Calculus does just that; it ignores the infinitesimals and assumes them to be zero. Although infinitesimals were ignored in calculus as a matter of convenience, we now discover that given the granularity of nature, such an assumption is necessary and accurate. In such a granular (rather than continuous) space, there would not be infinite number of points between two given points, rather the two points are separated by a very large (yet finite) number of points. The process of bringing the two points closer and closer would be limited by the minimum permissible dimension of space (Planck's length, or 1 unimeter, according to our discussion in the previous chapter). The two points cannot be brought any closer than this distance; any attempt to bring them closer would make them overlap.

Quantum mathematics: The new math consistent with granularity of nature. Revisiting the number theory and geometry

The number system started with measurement of natural phenomena. In its simplest form, it started with counting. These were designated as natural numbers (1, 2, 3, 4, 5, and so on). Later developments required introduction of fractions and these definite fractions were termed as rational numbers. It was seen that there were some ratios which could not be expressed as definite fractions (for example, pi or π which is the ratio between circumference and diameter of a circle); such ratios were termed as irrational numbers. The irrational and rational numbers, combined together are real numbers. Further developments led to the introduction of imaginary numbers (the squares of which yield a negative number) and complex numbers (any combination of real and imaginary numbers). Zero had already been introduced to explain and simplify many features of the number theory. We have discussed about zero earlier, and we shall have a relook at this entity (zero) again later. All the numbers have their mirror images with zero as the centre, and these are negative numbers.

We have seen in the previous discussions on overtaking vehicles and Zeno's paradox that nature has put restriction to the minimum magnitude of measurement of distance and time. This concept can be extended to matter and energy too. Max Planck showed that energy can be emitted or absorbed only in discrete energy quanta, and Einstein extended this argument to light saying that light consists of quanta of energy travelling through space. Thus nature has put a restriction on the minimum magnitude of the measurements of space, time, matter and energy. This is what I mean by the discrete nature and the lack of continuity of these basic components of reality, viz space, time, mass and energy.

In the previous chapter, we derived the minimum permissible values of distance (space), time, mass and energy. We assumed a new unit for each of these basic components of the universe in such a way that such minimum quanta of these entities can be represented as 1 unigram for mass (matter), 1 unimeter for distance (space), 1 unisecond for time, and 1 uni-joule for energy. With a suitable choice of units, we can say that the minimum magnitude of each of these fundamental entities as

1. All other measurements must therefore be multiples of 1. A fraction between 1 and 2 cannot exist because it would imply 1 plus something smaller than 1, and nature does not allow any magnitude smaller than 1 in the unit system that we have just considered. When we develop a system of mathematics to describe the phenomena concerning these basic entities of the universe where the minimum allowable magnitude is chosen to be 1, the entire number system becomes simplified to natural numbers. This is so far measurement of an entity is concerned. Of course, when one starts computing ratio between the measurements of two or more similar entities, there will be requirement of fractions and rational numbers. The need for rational numbers also arises when we wish to measure phenomena that can be expressed in terms of interaction between two or more basic entities, like speed (distance per unit time), momentum (mass multiplied by speed) etc. For example, if an object travels 3 unimeter in 6 seconds, the speed of the object would be 3/6 or ½ or 0.5 unimeter per unisecond. This value of speed is a fraction. But when we consider each of the fundamental entities (space, time, matter, energy) separately, all measurements can be reduced to natural numbers if we perform the measurements using the units as described a little earlier (unimeter, unigram, unisecond, unijoule). When we measure in other units, these values can be expressed as fractions or rational numbers. Nowhere in the scheme of measurement of physical entities would there be a necessity of irrational numbers. For the ease of illustrating this point, let us consider measurement of distance (space), and we can apply the same to the measurement of mass/ energy as well as to time.

In terms of distance, there must be a minimum distance below which there cannot be any smaller distance. This simply means that two points in space can come close only till a limit; the limit of this proximity has been set by nature as the minimum possible distance (1 unimeter as we have discussed). The two points cannot come closer than this in reality. However, in the system of real numbers, between any two real numbers, no matter how close they are, there will always lie an infinite number of real numbers. This is similar to the concept of continuum wherein between any two points in space, howsoever close to each other, there will lie an infinite number of points. Now, when we start dividing a physical distance and keep on continuing as a matter of mathematical exercise, we can go on indefinitely to ridiculously small scales of distances. The question however is: is this mathematical exercise also a reflection of physical reality? We have considered the problem of continuum in the paradox of overtaking vehicles. My view on this is that the concept of continuum, as well as the concept of irrational numbers and real numbers, is mathematical idealization. In reality, there must be a minimum allowable distance (1 unimeter in our assumed unit for distance) below which the concept of distance will lose its meaning.

If we accept this view of a minimum permissible distance between two points, the space then is no longer a homogenous and a continuous entity. Rather space is in its intrinsic design, granular. For this concept to be integrated in mathematics, it requires a wide range of modifications in the most basic concepts of geometry and mathematics. Let us start with the concept of a circle. The whole idea of circle in geometry, since the most ancient times, has been that of a regular curve that closes on itself and each point on that curve is equidistant from a point within the confines of the curve, called as the centre of the circle. The curvature of the circle is supposed to be continuous. But should it really be so? Think of the points on the circumference of the circle. The idea of continuous curvature is directly linked to the idea of two adjacent points on the circle being infinitesimally close to each other. But according to our hypothesis of granularity of space, there must be a minimum permissible distance between two adjacent points on the circle's circumference. So when all the adjacent points on the circumference of the circle are connected, the circle turns out to be actually a polygon with extremely large number of sides (*Figure 13*). This irregularity in its curvature may not be very apparent even on very high magnification, but it must indeed be so if the hypothesis of granular space is to be believed.

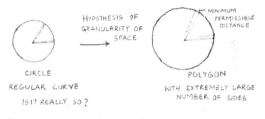

Figure 13: Is circle actually a polygon with an extremely large number of sides? The continuous curvature of a circle may not be true.

That the circle is nothing but a polygon with extremely large (not infinite) number of sides changes the calculations of the circumference of the circle. In this picture, the circumference of the circle is actually the sum of the lengths of each side of the polygon that the circle represents; that is the perimeter of that polygon. This value, whether we call it the circumference of the circle or the perimeter of the multi-faced polygon) must necessarily be a simple multiple of the minimum permissible length already discussed. The diameter of the circle should also be a simple multiple of the minimum permissible length. Therefore the ratio between the circumference and the diameter of the circle should be a simple fraction. But we know from our school days that the ratio between circumference and diameter of a circle is pi (π); and this π is an irrational number, not a fraction. The irrationality of π is because of the mathematically idealized concept of a circle whose curvature is continuous. But as we have seen, such an ideal circle with continuous curvature is inconsistent with the concept of granularity of space. As the hypothesis of granular space transforms a circle into a multi-faceted polygon, this irrational number π is simplified into a simple fraction.

The transformation of a circle to a polygon with numerous sides seems initially as just a subtle change, a mere playing with shapes. But we have seen how this has simplified the notion of π, transforming it from an elusive irrational number to a simple fraction. It is now also easy to explain the concept of tangent on a point on a circle. Imagine a stone being attached to a string and being moved in a circle by a child. When the child suddenly releases the thread from his hand, the stone flies off in a straight line from the point where it was when the child released the string. This straight line that the stone would follow is the tangent on that point on the circumference of the circle. But if the motion of the stone were truly regular on the curve without breaks, the stone should have followed the same curvilinear path. Instead it follows a straight line off the tangent. Why? It is so because the motion of the stone on the circle is not actually regular; rather it changes direction at every point on that circle; each point being separated from its adjacent point by a finite distance (the minimum permissible distance, say 1 unimeter). The child must hold the string and put some effort to hurl the stone in a circular fashion. The effort of the child provides the force required to change the direction of the stone at every point, even when it is moving at a constant speed. When the child releases the string, she also stops applying any force on the string and thus the stone would now follow the path in accordance with the law of inertia of motion. It would continue to move in a straight line in the direction determined by the line connecting the point at which the string was released and its immediately preceding point on the circle. This line is then the tangent on the point on that circle.

We shall consider a few more consequences of this concept in geometry. If we draw two lines joining the centre of a circle to two points on the circumference, these two lines are at an angle with each other. Now imagine this angle to get smaller as the two points on the circumference move closer to each other. If space is granular and if there must be a minimum permissible distance between two adjacent points, it also follows that there must be a minimum permissible angle between these two lines if these two lines are not superimposed. This angle would simply be the angle formed by the two adjacent points that form a side of the multi-sided polygon representing the circle to the centre of that circle. Bigger the circle, farther away would be the circumference from the centre; and thus the angle made by adjacent points on the circumference to the centre would be smaller for the bigger circle than that for a smaller one. This minimum permissible angle would vary from one circle to another, but clearly there is granularity even in the angles. The transition of angles is not smooth and continuous from zero degree to 360 degrees in a circle; this transition is in small jerks, allowing some angles and not allowing others. Granularity of angles is a direct consequence of the granularity of space. We shall see now that this granularity of angles has the potential to change many other long-standing mathematical notions.

We shall consider Pythagorean Theorem now. Nearly 3000 years back, the Greek mathematician and philosopher Pythagoras proved that the square of the hypotenuse of a right angled triangle is equal to the sum of the squares of the length of the other two sides (base and perpendicular) of the triangle. This theorem was discovered by an Indian mathematician Baudhayana a few centuries before Pythagoras, and the Chinese too knew about it before Pythagoras. But the theorem carries the name of Pythagoras, probably because of the dominance of the Greek philosophers, scientists and mathematicians in shaping the western scientific temper. According to this theorem, in a right angled triangle,

$$(\text{Hypotenuse})^2 = (\text{Base})^2 + (\text{Perpendicular})^2$$

So, if the base and perpendicular of a right angled triangle is 1cm each, the square of the hypotenuse should be 2 cm^2, and therefore the hypotenuse should be $\sqrt{2}$cm which happens to be an irrational number, that is, it cannot be expressed in terms of a simple ratio of two natural numbers. Remember we had raised objection to the concept of irrational numbers just as we had objected to the concept of continuum. Let us consider this problem more analytically. The entire premise of Pythagorean Theorem is based on a right angled triangle, and therefore on the fact that right angle (90^0) does truly exist. In our preceding discussion, we have seen how the concept of uniform curvature of a circle is fallacious. Rather, the circumference of a circle is actually comprised of several small straight lines of 1 unimeter length. Therefore a continuum of angles is also not possible. Only those angles would be permissible by nature which can allow the sides of a triangle (or for that matter a hexagon or a pentagon or any such shape) to have a length that is a simple multiple of the smallest basic length of 1 unimeter. The angles between the sides of the triangle (or any other shape) will accordingly be determined. So, just as there cannot be a perfect circle or sphere in nature, there cannot similarly be a perfect right angle. The lack of my formal mathematical training beyond school becomes a limitation for me to take this further and formulate this system of math and make it more general. I sincerely hope that some mathematician may, one day, describe the exact execution of this principle.

We have seen so far that for the purpose of measurement of distance, the number system can be reduced to that of natural numbers if a suitable unit system is used (in our discussion we have used the unit system of unimeter, 1 unimeter being the smallest distance possible). For the purpose of comparisons of distances and ratio between them, rational numbers may need to be brought in. So, what about the system of real numbers? This brings us to the discussion about mathematics in its pure form and its application to the physical sciences and the actual reality. Mathematics in itself is a wonderful discipline. The elegance, harmony, power and consistency of mathematics is incomparable to any other discipline. It has its application in almost all the study disciplines in science and arts. Particularly for the physical sciences, mathematics forms the backbone of the laws of nature. Indeed, it can be said that mathematics perhaps is closest to the language of Gods. However, in the context of application of mathematics to physical sciences, I think that both mathematicians and physicists have gone overboard. In place of this mathematics based on continuum, we have to consider this new math; I would call it Quantum

Mathematics that would be consistent with the granular nature of reality. I shall attempt to justify this statement now.

If we consider the system of real numbers, we assume that these numbers provide the magnitude needed for the measurement of different physical entities like distance, mass, energy, time, and so on. Many of the discussions in modern physics are based on this assumption. In fact, the number system was extended beyond the real numbers by the introduction of imaginary numbers and complex numbers; these too found their application in physics. For the sake of clarity and simplicity, let us confine ourselves for the time being to the system of real numbers and its application to physical universe. Is there a corresponding physical reality for each number in the real number system? To me, the answer is no. The system of real number is based on the system of continuum, and we have already discussed about the problems of the concept of continuum. If the concept of continuum is correct and is to be believed in physics, then there would be no limit to which a particular distance can be divided, there would be no limit to the minimum possible magnitude of distance, and no vehicle could ever overtake another vehicle (see our discussion on the problem of overtaking vehicles). This clearly implies that real numbers are mathematical idealization and we should not be overzealous in applying this system of real numbers to actual physically objective entities. In geometry too, we need to modify certain concepts. As we have discussed, the concept of uniform curvature of a circle and that of continuum of angles will need modifications. Of course, these concepts work beautifully in day to day world. But at really small scales that approach the minimum permissible distance (1 unimeter), such uniformity may not exist.

The preceding discussion deals primarily with distance (a property of space). We can use similar logic and reasoning for the other physical entities of time, mass, energy, and so on. Hence a particle cannot be reduced below a certain minimum permissible discrete mass. The duration of time cannot be shorter than a certain minimum permissible discrete time interval. And there cannot be an amount of energy less than a minimum quantum. Nature thus is not a continuum; rather it is granular and composed of discrete entities.

14
Infinite instability of zero and creation of the universe

"Zero is powerful because it is infinity's twin. They are equal and opposite, yin and yang. They are equally paradoxical and troubling. The biggest questions in science and religion are about nothingness and eternity, the void and the infinite, zero and infinity. The clashes over zero were the battles that shook the foundations of philosophy, of science, of mathematics, and of religion. Underneath every revolution lay a zero – and an infinity."

■ Charles Seife (Zero:
The biography of a dangerous idea)

When I presented the "split of zero" hypothesis about the origin of the universe, we also discussed about the problems that this theory faced (see Chapter 9). Although "split of zero" can be an esthetic explanation about the origin of the universe, it could not explain how the positives and negatives in the universe were compartmentalized. It is possible that positive space, positive time, positive mass and positive energy; and negative space, negative time, negative mass and negative energy – these exist in separate universes that are separated by an unknown dimension. We may be living in one of the positive universes, and experience the world around us the way we do. In the negative universe, the arrow of time may be reverse, that is time may run from future through present to the past. The phenomena of motion, gravity etc. may exist in similar way or in a much different way in the negative universes. If a connection between a positive and a negative universe is established through the weird unknown dimension that separates them, the two universes might interact to actually annihilate each other; negatives and positives combining to result in zero or nothingness. While these are great ideas of science fiction, they do not seem appealing enough to me. Although I do not completely discount such a possibility, I would personally favor another explanation of the split of zero into negative and positive entities.

I have discussed about this alternative explanation earlier in this book; briefly this explanation is as follows. The basic entities that we discussed – space, time, matter and energy – might be inter-related in such a way that their sum total is zero. These entities are born out of nothingness and assume such different characteristics that it seems difficult to imagine that they were once born of the same parent (nothingness). Space, time, matter and energy become so much specialized and differentiated entities that they cease to interact to get back to their sum total of zero (nothingness). These basic entities do interact with one another, but this interaction doesn't result in their cancelling one another. Rather, their interactions instead result in the other phenomena like motion, speed, momentum, density, force etc. Thus space and time would interact to give rise to the phenomenon of speed or velocity, matter (mass) and velocity would give rise to the momentum. Matter, distance, time and energy may interact in other different ways to give rise to force, gravity, and other phenomena of nature.

Another equally or perhaps more important problem with the hypothesis of "split of zero" is that it could not answer why other objects do not split spontaneously. What is so special about zero that it could split to give rise to the universe while the other objects (say a table or a chair) do not split in our day-to-day experience? If nothingness (zero) was self-sufficient, why did the split happen? Why is there something rather than nothing? We shall enquire into this; the special nature of zero and the reason behind the primal split of nothingness.

The contradictions shown above are the major limiting factor of the theory of "Split of zero." In order to resolve this problem, we need to study the nature of zero. For this, I would put forward a hypothesis and then we shall try to support the hypothesis with logical reasoning and available data.

Hypothesis: "The stability of a particular entity is inversely proportional to its proximity to zero."

By "entities," we mean mass or energy or space or time or any other conceivable phenomena in the universe.

This would mean that gross entities would be stable because they are quite distant from zero. However, the smaller it becomes, more is its instability. The closer it gets to zero, its instability would increase to such an extent that it can never become zero. In other words, zero would be infinitely unstable. Therefore, zero (nothing) cannot exist by itself. The only way how zero can exist is in the form of positives and negatives, the sum total of which is zero (*Figure 14*).

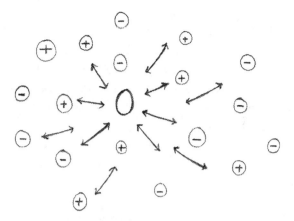

Figure 14: Infinite instability of zero: Zero or void cannot exist on its own; rather it exists in the form of myriads of entities and phenomena in the universe whose sum total must be zero.

Is this hypothesis supported by available observations and data? We have seen that the major objection to the "split of zero" hypothesis was that if zero can split, why the other objects do not split. In the day-to-day macroscopic world, we see that the things are relatively stable despite the internal flux and dynamics at their molecular, atomic and subatomic level. A table or a chair in a room would remain as it is unless there is an external disturbance or force acting upon it. This stability of the entities and the resultant predictability of their motion and behavior also apply to the phenomena seen in the universe at a comparatively larger scale that is quite distant from zero. The course and motion of a soccer ball can be predicted with accuracy, provided the force and the direction of the kick, the speed and direction of the wind, and the airway resistance are known. Similarly, one can accurately predict the motion of the planets and the constellations. With the help of these calculations, we can predict eclipses and other phenomena.

Things start becoming quite different as we move to smaller and smaller scales that are closer to zero. At the subatomic level, uncertainty creeps in. The position and momentum of an electron cannot be determined with certainty. Elements of chance and probability dominate at quantum levels. The photon or an electron appears to have a whole range of probabilities of its motion while passing through a slit. The double-slit experiment is one of the cardinal experiments supporting the quantum theory. We shall elaborate on the details of this experiment later when we discuss about the quantum theory in the light of my supposition of a granular universe. The strange inference drawn from the double slit experiment is that in addition to the wave-particle duality of the photon or electron, the photon or the electron somehow seems to know if the other slit is open or not, or whether there is an observer detecting its passage through one slit or the other. This rather extraordinary phenomenon of interaction between the observer and the observed shows quite clearly that there is a considerable degree of instability, uncertainty and unpredictability at such micro scales.

While discussing the Big-Bang theory, we saw that the evolution of the universe could be explained fairly accurately from the stage of the primordial fireball provided we did not question what happened within the first 10^{-43} seconds of the actual Big-Bang. The laws of physics break down at the point of singularity where the density, temperature and curvature of the extremely dense matter-energy complex were close to infinite. The point of singularity again corresponds to an extremely small scale. The entire space, matter and energy of the universe was then concentrated in an extremely small region. The duration of 10^{-43} seconds after the Big-Bang is again an extremely small duration of time. When such small scales of space and time are under consideration, the laws of physics break down and things are extremely unstable and uncertain. It is conceivable that what we call as singularity actually represents such a close proximity to zero that space, time, matter, energy etc. lose their identity and distinction from one another. This is my idea of singularity: a situation so close to zero that all the basic entities that define the universe lose their individual identities. So great would the instability be of such "singularity" that it would transform itself to a myriad of entities like space, time, matter, energy and so on. These entities and their interactions would then give rise to the entire universe as we know it. The stability of these entities and our ability to predict increase as we gradually move away from zero to the macroscopic scale.

The nature of zero and the secret of the origin of universe

Zero thus is a virtual reality. It is a reality because it is the sum total of the entire contents of the universe (including space, time, matter, energy, etc.). It is virtual because it cannot exist on its own. We have discussed the hypothesis of increasing instability of an entity as it becomes smaller and smaller, that is comes closer and closer to zero. By this reasoning, zero would be infinitely unstable. Infinite instability is equivalent to the impossibility of independent existence. This implies that zero cannot exist on its own, it can exist only as a multitude of entities and phenomena in the universe so long that the sum total of everything remains zero. In other words,

(Space + Time + Matter + Energy) of the entire universe = Zero (0)

Thus, the question about the origin of the universe does not arise. It cannot have a point of origin because the origin must come from zero or nothingness, and nothingness cannot exist. The universe exists because nothingness cannot. It does not require to have originated; rather it could not have had an origin. Thus, the universe never began, nor will it ever end. It has and will always exist as an infinite multitude of positive and negative things and phenomena, the sum total of which will always be zero.

This suggestion does not rule out a Big-Bang. It only says that the Big-Bang cannot be the origin of the universe. It might well have occurred, and if so, the evolution of the universe has been a consequence of the Big-Bang. The nature of the singularity that gives rise to the Big-Bang can be understood if one considers it as something that has come so critically close to zero that it becomes extremely unstable. This extreme instability of the so-called singularity results in the splitting of the singularity. This gives the appearance of an extremely hot and violent explosion, the so-called "Big-Bang." Several such big and small bangs might have occurred and might still be occurring. We may not have the capability of detecting and describing them. But, in principle, if anything is brought critically close to zero, it will split. This splitting may manifest as an uncertainty in its position and momentum. It may result in the particular entity to have a dual nature of a particle and a wave. It may result in the rules of probability that govern the quantum world. Occasionally, the splitting may be really grand and result in the formation of space and time, matter and energy, and give rise to the universe that we live in.

There may or may not be other kinds of universes. If there are, they may well have absolutely different characteristics. Our universe is characterized by the basic entities that comprise it: space, time, matter and energy. Some other universe may have different kinds of entities that we may not be able to comprehend. If there are multiple universes indeed, the sum total of all the entities in all the universes combined together must be zero.

It is this fluctuation around zero (nothingness) that creates space, time, matter, energy and whatever else we can perceive or imagine. Zero is the undifferentiated primordial source of all the entities and phenomena of the universe and yet, it cannot exist by itself. What exists is the conglomerate of space, time, matter, energy and so on. The universe manifests as these entities become differentiated and become somewhat stable. We can draw an analogy with the biological science, although it may not be an accurate comparison. When a sperm fertilizes an ovum, the result is a single cell that is undifferentiated and has within it the potential to form the entire human body with all its complexities. When we look at a human baby, it is a miraculous product of the unicellular fertilized ovum. The hands and feet, the eyes and ears, the heart and brain have become so much differentiated and specialized that it is difficult to conceptualize that all these owe their origin ultimately from a single cell. Similarly, it is difficult to conceive that the attributes of a large tree with its strong roots and trunk, its leaves and branches, its flowers and fruits were all once contained within a single seed.

A similar miracle works in the very fabric of reality. Zero or nothingness is that undifferentiated whole with the potential of the entire universe. Space, time, matter and energy are the differentiated products of this undifferentiated nothingness. When we think of these entities, it is difficult to believe that they might have a common origin, a common substratum.

But once we accept this, it opens an entirely new approach to explaining the nature of reality.

Thus zero (or nothingness) is the abstract substratum of the whole universe; it is the primordial undifferentiated whole and sum total of all that exists including space, time, matter, energy and other phenomena. However, because of its infinite instability, zero cannot exist on its own and must necessarily manifest in the form of myriads of entities and phenomena that form the universe. The universe exists simply because nothingness can't. Zero is the unmanifested whole; universe is the embodiment of its manifestation. Universe is what we perceive the reality to be; zero is the reality in its most naked form. Zero is the primal cause; universe is the result of the infinite instability of zero.

15
Causality and free will

"Cause and effect, means and ends, seed and fruit cannot be severed; for the effect already blooms in the cause, the end preexists in the means, the fruit in the seed."

■ Ralph Waldo Emerson

"Man can do what he wills, but he cannot will what he wills."

■ Arthur Schopenhauer

Our previous discussion on the special theory of relativity has taught us that nothing can exceed the speed of light. We also discussed about the concepts of length contraction and time dilation at speeds comparable to the speed of light. An object travelling at such remarkable speeds increases in its mass, contracts in length and experiences a slower transit of time.

Let us imagine what happens at the speed of light. We know that light is made up of photons. Let us ignore for the moment the controversies regarding the dual nature of light (particle and wave) and focus ourselves at what each photon would experience. According to the laws of special relativity, any particle with a rest mass cannot be accelerated to speed of light because the mass of the particle would then approach infinity and would require infinite force to be accelerated to speed of light. Hence, the photons that travel at the speed of light are considered massless (with no rest mass). Because of the principles of length contraction and time dilation as discussed earlier, space would contract infinitely and time would dilate infinitely for such photon. In other words, photons would also have no space dimension, and would not experience the passage of time. From the point of view of that photon, space and time would not exist. We shall now focus on the passage of time from the point of view of a photon.

We know that light takes about 8 minutes from the sun to reach the earth. Alpha Centauri, the nearest star other than our sun is about 4 light years away, that is light takes 4 years to travel from Alpha Centauri to our planet. However, all these durations of time that we are talking about are the durations that we perceive on the earth. For the photon of light traveling from the sun or the Alpha Centauri to the earth, this duration is zero; it has instantaneously reached the earth from the sun or any other star. Why? Simply because special relativity teaches us that at the speed of light, the time dilates to infinity, and hence it would stop ticking. From the perspective of a photon, there is no such thing as the passage of time. The entire human civilization, from the Stone Age through the stage of cavemen, hunters, agriculture, industrialization, colonialization, leading to the modern scientific era, would be just a flash from the photon's perspective. The birth of the sun, the earth and other planets, the cooling of earth, the beginning of life, the evolution of man – all this has taken place in an instant for the photon moving at the speed of light.

If we were to extend this argument, the photon emitted at the moment of big bang would not have experienced any passage of time till today and also will not experience any passage of time in the future. The age of the universe, which might be billions of years, is zero for that photon. The entire drama of the universe that has happened so far since its origin would be experienced as simultaneous events from the point of view of that photon. In fact, even the future would be simultaneous with the present and the past. From the photon's point of view, there is no difference between past, present and future. All the events, creation and destruction of the universe are before it simultaneously. Perhaps God resides on the moving photon!

It is not just about the passage of time, distance too has no meaning from the photon's perspective. At the speed of light, the distance contracts by an order of infinity, and thus the moving photon experiences no distance. The distance between the sun and the earth, between the earth and the moon, between one galaxy and another – all these are measured from the perspective of an observer on the earth from his or her frame of reference. However, from the perspective of the moving photon, such distances are meaningless. For a photon that was emitted at the moment of the big bang, not just the age of the universe is zero but the entire expanse of the universe is confined to a point (without a space-dimension).

So, if the special theory of relativity is correct, the photon moving at the speed of light has zero rest mass, zero space dimensions, and zero passage of time. Past, present, and future are all simultaneous from the point of view of this photon. This brings into question the concept of causality. If there is no past, present and future for this photon, all our lives and the entire drama of the universe, including the future is a simultaneous flash of events from the photon's perspective. We as mortal beings perceive the passage of time and thus would regard an event in the past to be the cause of another event in the present, and events in the past and present to be the causes of another event that would take place in the future. We also feel that we can perhaps try and change the future by our action. But there is this photon that has seen it all – our past, our present and our future. What does it mean? Whether or not God resides on this photon moving at the speed of light, it shows that the difference between past, present and future is an illusion; that there is no such thing as causality. I hate to believe this, but going by this logic, we have really no free will. Everything in the universe is in that sense, predetermined and has already been experienced by this moving photon.

It was the discipline of astrology and palm reading that claimed that everything in one's life is predetermined. This kind of fatalistic attitude was supposed to be the attitude of the meek and the weak; this was the attitude that people with scientific temper sought to get rid of. This scientific temper has led to remarkable inventions and discoveries that have changed the face of human civilization. It has led us to betterment of our lives through the application of modern technology; it has also led us to the grim realization of the possibility of our own destruction through our self-created environmental crises and through mindless warfare. But whether it leads us to glory or to destruction, science has taught us to take responsibility of our own actions and our own conduct. We cannot hide behind the fatalistic attitude of the unscientific mind. And now, lo and behold! Look at the biggest irony of all! The theory of relativity that can be considered as one of the greatest scientific theories of all times leads us to the most extreme form of fatalism. The scientific temper within you and me rebels at this conclusion; there must be something missing, there must be something that is not entirely correct about it. The entire plot of the universe cannot just be a theatre with all events that are predetermined.

This is a scary thought and I have my own views to the contrary. This is a view that would require some modifications to the theory of special relativity. Recall our previous discussions about the granular nature of the universe and the minimum permissible values of distance, time, mass and energy. In my view, there must be a reason why speed of light should be the highest speed permitted by nature. **The lowest permissible time interval, lowest permissible space dimension (distance), lowest permissible mass, and highest permissible speed – all these must have a common central theme.** Simultaneity of past, present and future would mean that time has stopped (as we have seen in the preceding discussion about the photon travelling at the speed of light). This would mean that the time interval for this photon would become zero; similarly, rest mass and space dimension of the photon would also be zero. My position, on the contrary, is that neither of these measurements can be zero for the photon or for any other particle. According to my view, in contrast to the special theory of relativity, time would stop only for a particle that has the speed of infinity, and to reach this speed of infinity the particle should also have zero rest mass. But as we have seen, there are minimum values of time, distance, and mass that are permitted by nature and that these values cannot be zero. So, for the minimum permissible mass occupying minimum permissible space, there should be a maximum permissible speed. It so happens that the photon is such a candidate with the minimum permissible mass and maximum permissible speed. Photon, contrary to the present scientific belief, should not be mass-less. Rather the mass of a photon should be the minimum mass permitted by nature; this would in turn permit it to have the maximum permissible speed (the speed of light).

Durgatosh Pandey

Past, present and future therefore do exist and cannot be simultaneous. Causality and free will do exist. We still have hope that our actions can change the course of the future, just as the past has shaped our present. Of course, there is an element of unreliability, there will always be certain amount of unpredictability, or luck as some would like to believe. But the universe is not completely fatalistic. All is not pre-decided. The only exception to this would be something that moves with infinite speed, occupies zero space and have zero mass. We have seen from our discussions repeatedly that nature does not allow this. If God were to exist, He ought to be mass-less and be moving at an infinite speed. He would then be omnipresent and timeless in the real sense of these terms. He would also then be outside the purview of our direct understanding of nature.

16
Quantum theory explained

"The conception of objective reality... has thus evaporated... into the transparent clarity of mathematics."

■ Werner Heisenberg

What is reality? According to the view of classical physics, there is an objective world that is independent of an observer viewing it. The earth would keep on revolving around the sun, days and nights would continue alternating with each other, water would keep on flowing in the river towards the ocean, mountains would keep standing tall, stars would keep on appearing in the night sky; all these would keep on occurring whether or not we are there to observe it. This view of the reality is also the common-sense view. However, the subatomic particles behave in quite bizarre way that cannot be explained by classical view. Quantum theory presents an alternative view of reality. In this view, reality does not exist in the absence of an observer. There is no objective world independent of the observer. It is a shockingly uncomfortable and mysterious theory that also talks about "alternative histories" of a particle wherein it does not follow a single path between two points; it rather follows every possible path connecting these points, and it takes all these paths simultaneously. The behavior of a particle can be estimated only in a probabilistic fashion. The act of observation or measurement by an observer reduces these alternative histories into one particular path; the so-called "collapse of wave function." The probabilities are reduced to one reality due to the act of observation. Niels Bohr, one of the founders of Quantum theory, said, "Everything we call real is made of things that cannot be regarded as real… If quantum mechanics hasn't profoundly shocked you, you haven't understood it yet."

A full discussion on the nuances of quantum theory is outside the scope of this book. I have no specialist knowledge of quantum theory nor the understanding of its mathematical foundation. I have attempted to understand it in an amateur fashion in order to satisfy my curiosity that has led to so many streams of knowledge, including quantum theory. With this disclaimer, I shall try to deal with a few aspects of this theory and attempt to explain them using my hypothesis of impossibility of a continuum and the discrete and granular (rather than continuous) nature of space and time. Three fundamental concepts in quantum theory that will be discussed include:

1. Dual nature of light (wave as well as particle)
2. Uncertainty principle of Heisenberg
3. Double-slit experiment

We shall discuss these three concepts one by one.

Dual nature of light

Is light a wave, or is it composed of particles? The debate went on for centuries. Newton considered light to be made up of a stream of tiny particles called corpuscles. He could explain the behavior of light and the concepts in optics using this theory. For instance, light travels in a straight line just as a particle would do. The reflection of the light from a mirror is akin to the bouncing back of a ball on hitting a wall. Despite the dominance of this theory of the particle nature of light, an alternative view has been there even during Newton's time. Christiaan Huygens, who was Newton's contemporary, believed that light is a wave. Just as the waves move across the surface of a pond or a sea, light waves were believed to propagate through an all-pervading invisible substance, the mysterious so-called luminiferous ether.

It is now believed that light has a dual nature. It is composed of photons, the particles or corpuscles of light. It also has a wave-like character. Each photon, for that matter, has a dual character of particle as well as wave. The dual nature does not confine to photons; it also applies to electrons, other subatomic particles, and to every conceivable particles and objects in the universe. How do we explain the particle behaving like a wave?

Imagine that you are watching a Formula One race. Before the race, the cars are at rest and are not moving with respect to the spectators of the race. It is very easy to take a good still-photograph of the cars at rest. As the race begins, the cars speed up. If you try to take a still-photograph of a moving car, the picture would be a bit hazy. The faster the car, the hazier the picture would be. At very high speeds, the picture of the cars may be smudged. Similar is the case of particles at quantum scale. We shall return to the photon for the sake of simplicity; the same argument would apply for all the particles and objects. The photons are the corpuscles or particles of light, yet the speed at which they move make them appear to be wave-like. One should also appreciate that the particle, photon, is a packet or bundle of energy carried by the light.

The smudging of the photon at the speed of light causes this energy to be spread out and this gives the photon a character of a wave. This applies to the electrons and other subatomic particles too. There is a difference, however, between the waviness of particles at quantum scale and the haziness of the photograph of the fast moving cars. While the haziness of the picture of a moving car is because still-photography is not a good technique to follow the car in motion, the waviness of the particles in quantum scale is because of the spreading out of the energy associated with it. This spreading out of the associated energy is an intrinsic property of the particle. Therefore, waviness is an inherent nature of a particle.

In the quantum world, the wave aspect of a photon or an electron can be viewed as a continuous spread over space and time, while the particle aspect can be considered as a localized entity within space and time. As we have discussed earlier, reality is granular and discrete; and Nature abhors continuity. As soon as one considers a point-like localization of a moving electron or a photon, and we zero-in on the location of the particle, its location becomes unstable because of the principle of instability of zero. Absolute localization of a quantum particle means allotting definite co-ordinates in space and time at that particular instant. The more pointed the localization, the greater becomes the instability of the particle and its tendency to spread out. Thus, wave nature of the quantum particle is the tendency of spreading out of a localized particle due to its inherent instability; and the particle nature of a quantum wave is a tendency of the wave to avoid continuity in space and time. The dual nature of the electrons and photons can thus be considered as an attempt to balance between the instability of absolute localization of a particle and the necessity of avoiding the continuity of a wave.

Does this also apply to the macroscopic world? To the cars and the airplanes, to the books and the buildings, to you and me? In principle, it does apply to the objects in our day-to-day life, but the speeds at which they move are far less than the scales nearing the speed of light. In such situations, the smudging of the picture is much less and the wave-like character of the macroscopic particles is not appreciable.

Uncertainty principle

Imagine a soccer player taking a free-hit. He kicks the ball which takes a curve above the defenders, dodges the goalkeeper, and enters the goalpost hitting the net behind. He has scored a goal! Let us now consider the motion of the football from the time that it is hit by the player till the time it hits the net. We can define the motion of the football accurately by knowing the force with which it was hit, the direction of hit, the speed of the wind (if any) and the air-resistance. Provided we know these variables accurately, we can predict exactly where its position and velocity would be at a given time. But, does such a deterministic principle apply in the quantum world? Werner Heisenberg discovered that these familiar rules of everyday world do not hold in the subatomic micro-world of electrons and photons, rather they are governed by uncertainty relations. The precise position and momentum of a particle cannot be determined accurately at a given time. The more accurately we know about the position of a particle, the less accurately can we determine its momentum, and vice versa.

Initially, this uncertainty principle was thought to be due to the difficulties in the act and process of measurement in the context of the subatomic particles. We can see an object only by looking at them, which involves the rays of light (consisting of photons) striking the object and bouncing back to be detected by our eyes or by a detector. These photons do not have enough energy to disturb the football's motion and hence the position and velocity of the football can be determined with a great deal of accuracy. On the contrary, if we consider the motion of an electron, things are quite different. The detection of an electron requires gamma radiation that is quite energetic. A photon of the gamma radiation that hits the electron and bounces off to be detected by our detector will drastically change the position and momentum of the electron in question. The very act of observing an electron has created this uncertainty in determining its position and momentum.

Is uncertainty then because of the inaccuracies on account of errors in measurement? No, uncertainty in the micro-world is a matter of principle and is not just due to the interference by the act of measurement. According to the fundamental principles of the quantum theory, an electron cannot possess both a precise position and a precise momentum. Heisenberg wrote: "We cannot know, as a matter of principle, the present in all its details." This is a significant departure from the determinacy of classical physics. According to the classical physics described by Newton, future can be predicted accurately if we knew the position and momentum of every particle in the universe at a given time. This becomes meaningless in the context of quantum theory, according to which no particle possesses a precise position and a precise momentum. The uncertainty principle declares that the uncertainty in position and momentum is an inherent character of a particle, and is not just a consequence of the act of measurement. This uncertainty principle of Heisenberg is the central feature of quantum theory.

If we denote the momentum of an electron by p, and its position by q, there would be some "error" (due to uncertainty) in the measurement of these characteristics of the electron. The error in measurement of momentum can be denoted by Δp and that of position by Δq. According to the uncertainty principle, there is an absolute minimum limit to the product of the two "errors". Mathematically,

$$\Delta p \times \Delta q \geq \hbar/2,$$

where $\hbar$ is the reduced Planck's constant (h), and $\hbar = h/2\pi$.

Planck's constant (h) is the proportional constant between the energy (E) of a photon and the frequency (f) of its associated electromagnetic wave [E = hf].

Please do not get intimidated by the equations; it is the principle that is much more important. What does all this mean to a lay but curious and interested person? Very significantly, it means that even if the tools of measurement of position and momentum of a particle (say electron or a photon) be refined to a degree where the errors are no longer due to measurement, the uncertainty factor will introduce these errors. Uncertainty, therefore, is a fundamental property of the particles in quantum world, and is independent of measurement. Also, the more accurate the position of a particle, the greater is the inaccuracy or uncertainty in its momentum, and vice-versa.

How can this uncertainty be explained? Are the position and momentum of a particle really a matter of chance? Are the events really governed by probability? Is God playing dice in this universe? Let us see if we can explain the principle of uncertainty to be a direct consequence of the impossibility of continuum of space and time. My only assumption is that continuum is not tenable in nature, and there is a discrete minimum value of space (distance) and a discrete minimum value of time duration allowed by nature. The logical and mathematical basis of this assumption has already been dealt with previously.

In the previous discussions, we have seen that nature allows a minimum quantum of length (Planck length, or 1 unimeter with our new measurement units) and a minimum quantum of time duration (Planck time, or 1 unisecond with our new measurement unit) lower than which any magnitude of length or time is not permissible. Similarly, we have seen that there must be a minimum quantum of mass and energy permissible by nature. But presently, we will discuss a bit more about space (length as one of its dimensions) and time, and see how the principle of uncertainty can be explained.

The crucial common factor in the relationship between Planck's length (1 unimeter) and Planck's time (1 unisecond) is the speed of light in vacuum (c). The duration required for light to cover a distance of 1 Planck length is 1 Planck time. Thus, a photon travelling at the speed of light travels 1 unimeter in 1 unisecond. Thus speed of light in vacuum in this unit system is 1 unimeter per unisecond.

C = 1 unimeter/unisecond (1 um/us).

We know from the special relativity theory of Einstein that the speed of light in vacuum (c) is the maximum speed allowed by nature. Any particle can travel at a speed that is lower than or at the most equal to the speed of light. The speed of 1 unimeter per unisecond is the maximum speed allowed by nature since it is the speed of light in vacuum. Let us now consider the motion of four particles A, B, C and D.

Particle A travels at the speed of light c (1 unimeter per unisecond).

Particle B travels at a speed of c/2 (1 unimeter in 2 uniseconds).

Particle C travels at a speed of c/100 (1 unimeter in 100 uniseconds).

Particle D travels at a speed of $c/10^8$ (1 unimeter in 10^8 uniseconds) = 3 meters per second.

The particle D moves the minimum allowable distance of 1 unimeter only after 10^8 uniseconds have passed, its position remains more or less well-defined for the time duration of 10^8 uniseconds. The position of the particle C remains more or less well-defined for a period of 100 uniseconds and the position of the particle B remains well-defined only for 2 uniseconds. The particle A remains in a given position only for 1 unisecond. What does this imply? To an observer who is trying to locate these particles in motion, it is easier to locate particle D at a particular position than the particle C, it is easier to locate particle C than B, and it is virtually impossible to locate precisely the position of the particle A that travels at the speed of light.

Can we relate this illustration to the principle of uncertainty? Every "jump" of 1 unimeter distance produces change/ uncertainty in the position of the moving particle. The slower the particle, the greater is the time duration for which it can be detected at a particular position. The faster the particle, the lower is the time duration for which it can be detected at a particular position. It is clear from this illustration that as the speed of a particle increases, the uncertainty of its position also keeps on increasing. In the quantum world, the speeds at which particles travel are quite high. As these speeds approach the speed of light, the uncertainty of the position of the particles involved increases.

Double-slit experiment

This is an experiment that touches the very nerve of the quantum theory. It has shown remarkably consistent results when performed by different physicists in different places with different particles like photons or electrons. Yet, equally remarkable is the fact that it doesn't make any sense at all; the particles of the quantum world behave differently while being observed than they do when they are not being observed. Let us examine this fascinating experiment in the layperson's language in a little detail.

Let us imagine firing bullets at a wall that has two holes or slits in it. Most of the bullets will collide with the wall and will not be of any consequence to us, but a few of them would escape through the slits in the wall to hit another wall behind. It is these escaping bullets that we are interested in. What then would be the pattern of the bullets striking the wall behind the slits? Clearly, there would be two linear bands on which all the bullets that have escaped through either of these slits strike. A bullet would either pass through the slit 1 and strike at band A of the wall, or would pass through slit 2 and strike at band B of the wall.

If we change the bullets with water ripples or wave, what would be the pattern on the wall behind the holes? This time, the water wave would strike the two slits and independent waves emerge through the two slits. These independent waves interact or interfere with each other and produce a pattern that is scientifically termed as diffraction or interference pattern on the wall behind the slits. When the crest of wave 1 coincides with the crest of wave 2 at the wall behind the slits, the result is a greater crest of twice the magnitude of each wave (constructive interference); and when the crest of one wave coincides with the trough of another at the wall, the result is cancelling of the wave (destructive interference); and there are interactions between these two extremes.

Now, let us see what happens when we substitute the bullets or water waves with photons or electrons (*Figure 15, 16*). We have seen that light has a dual nature; it behaves both as particle and wave. Electrons similarly have a dual nature of particle and wave. In the experimental setup now, we have electrons being fired from a source to a screen with two slits, and there is a fluorescent screen on which the electrons that come out through the slits would impact and produce a pattern of distribution recorded on the fluorescent screen. When one of the two slits is open and the other closed, the electrons have only one slit to pass through, and all these electrons are concentrated on a linear band on the fluorescent screen behind, just as the pattern that the bullets would have created on the wall behind the slit. However, when both the slits are open, the electrons behave like the water waves and there is a pattern of diffraction on the fluorescent screen.

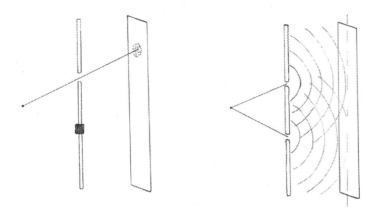

Figure 15: Double-slit experiment: Electrons fired from an electron-gun and escaping the slits behave like particles when only one slit is open and as waves when both the slits are open.

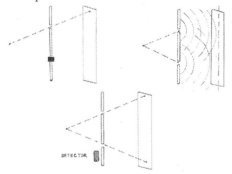

Figure 16: Double-slit experiment: Even when only a single electron is fired at a time, it behaves as a particle when one slit is open and as a wave when both slits are open; passing through both slits and interfering with itself. Quite amazingly, when a detector is placed to peep into the experiment to see which slit it actually passes through when both slits are open, the electron reverts to a particle-like behaviour.

We see therefore that the electrons behave as a particle when one slit is open and as a wave when both the slits are open. But where is the weirdness in all this? We know that photons and electrons have dual nature; they are supposed to behave both like particles and waves, and they are just doing so. The real problem is when the speed of firing the electrons is reduced to such an extent that only one electron is fired at a time. Now, this particular electron has either of the two slits to pass through. We decide to let this continue for some time and then observe the fluorescent screen for patterns. Common sense would suggest that when only one electron is fired at a time, it passes through either one of the two slits and impacts on the fluorescent screen at a corresponding point. Hence, after some time, we should see two bands on the fluorescent screen corresponding to the two slits through which individual electrons passed. In reality, however, the pattern that is observed on the fluorescent screen is that of diffraction, the interference pattern of a wave. How can this happen? Does it mean that one particular electron passes through both slits and interferes with itself while striking on the fluorescent screen? Yes, says the quantum theory. The electron is not a defined entity localized at one place; it is a probability wave and thus can pass through both the slits and interfere with itself.

Can this be explained by our hypothesis of impossibility of continuum and the nature being discrete? Remember that with this hypothesis, we explained that the particle cannot have a point location; its associated energy is necessarily spread out. When a particle like an electron has only one slit to pass through, the entire energy associated with that particular electron has no choice but to go through the open slit. However when both slits are open, the energy associated with a single electron is spread out, and can behave like a wave passing through both the slits simultaneously and interfering with itself after passing through the slits.

But now comes the most shocking part. Some scientists were not convinced that a single electron could pass through both the slits simultaneously and decided to peek into the experimental setup to see which slit an individual electron actually travels through. So, a detector was placed adjacent to one of the slits that would detect the electron while passing through it. When the electron was being observed, it started behaving like a particle and produced a pattern of two bands on the fluorescent screen; the interference or diffraction pattern disappeared. The very act of observing changed the way electrons behave; when not observed they behaved like waves, and when observed they behaved like a particle. The act of observation, it was concluded, collapsed the probability wave into a particle. The physicists were completely baffled, and they undertook one further step in this experiment. They suspected that the act of measurement or observation interfered with the motion of the electron, and this might perhaps explain the change in behavior from a wave to a particle. So the electron detector was kept near a slit, but no measurements were made, no data was collected. The detector was completely functional; the only difference was that data was not being collected, measurements were not being made. Now, the electron went back behaving like a wave; the pattern on the fluorescent screen became that of interference or diffraction.

This was crazy. It was as if the electrons knew that they were being observed; and not just that, they also knew when they were observed passively (without making measurements) or actively (while making measurements). This could not be explained by any possible means of rational explanation. The quantum world is far more mysterious than the scientists would have imagined. How does the particle or the wave in the quantum world know that it is being observed, and whether such an observation is associated with collecting data or making measurements? Just as schoolboys who would be all around doing all sorts of things in the absence of the teacher but will sit inside the class in an orderly manner when the teacher is around, the electrons move around in all sorts of funny ways interacting with itself and other electrons till an observer steps in; it then resorts to its orderly behavior of a particle. The conclusion is inescapable: there is something in the quantum world that interacts with the observer and the act of observation. The act of observation requires awareness and consciousness. No longer can the subject of consciousness be hushed in the discussion of physics. The discussion must come out in the open; the interaction between the observer, the observed, and the act of observation is the key to drama happening in the quantum world.

17

Consciousness: another basic component of the universe

"Science cannot solve the ultimate mystery of nature because, in the last analysis, we ourselves are a part of the mystery that we are trying to solve."

■ Max Planck

"Consciousness is the fundamental thing in existence. It is the energy, the motion, the movement of consciousness that creates the universe and all that is in it. The microcosm and macrocosm are nothing but consciousness arranging itself."

■ Sri Aurobindo

In the earlier chapters, I had made a passing mention of consciousness while discussing about the basic components of the universe. We discussed about space and time, matter and energy, but skipped consciousness; partly because it is a more difficult topic if discussed purely within the realms of science, and partly also because it would have been premature to introduce the topic of consciousness then purely as an abstract concept bordering on spirituality and theology. That has not been the purpose of this book. Now, the context is there and we can introduce this topic. We have seen the weirdness of the quantum theory especially that of the double-slit experiment where the act of measuring changes the way a particle (say an electron) behaves. We shall now discuss something about consciousness in general, and then try to see if consciousness as a basic entity of the universe can explain the weirdness of quantum mechanics.

Consciousness is the spark of life in every living being, the entity that makes an individual what he or she is. It is the "I" in me and "You" in you. It is the awareness of one's own self. Consciousness is too familiar as everyone has it; yet it evades an accurate definition. Consciousness is the most fundamental aspect of the self; it is the ultimate doer of all actions, as well as the ultimate perceiver of all external and internal stimuli. It is important to understand the nature of consciousness because it is the cause of all good and all evil. The ancient Indian philosophical text, Brihadaranyaka Upanishad says, "You cannot see the seer of seeing, you cannot hear the hearer of hearing, you cannot think of the thinker of thinking, you cannot know the knower of knowing. This is your Self that is within all; everything else but this is perishable."

There have been attempts to understand consciousness through scientific methods. The scientists have tried to explain it as a part of or the function of the brain. According to this view, consciousness could be explained as a composite of complex chemical reactions and electrical interactions in the neural network of the brain. While there may indeed be some merit in this explanation, there are many lacunae in it. This may explain how we perceive the outside world and respond to external stimuli. But, it does not explain who actually perceives it. In other words, it explains the action but not the doer of the action. This view would also mean the possibility, in principle, of an artificial intelligence with completely human potentials including awareness and consciousness. It is something that, in principle, doesn't seem agreeable to me.

The subject of consciousness has been discussed in religion, theology, metaphysics and psychology; but it has not received its due attention by science. If we have to study consciousness in a scientific manner, some of the rigid constraints of the scientific method will prove to be the limiting factors for such an endeavor. Science demands observational validity for a theory to be accepted. The problem with this requirement of the scientific method is that all observations must necessarily then be processed through the human brain and intelligence and finally perceived by the consciousness. Only then, we can know what has been observed. But in the search of the nature of consciousness, how can the observer be observed? Let me explain this a little more simplistically. We see through our eyes. However, to see one's own eyes, one must use a mirror. Thus, what we see is the image of the eyes and not the eyes themselves. Imagine, if the eyes that are just tools for the act of seeing cannot be directly observed, how can consciousness, which is the ultimate doer and perceiver, be directly perceived? Thus the observational methods are not the correct methods of understanding consciousness. Such concepts do not necessarily yield to the requirements of scientific methods of known facts and reproducible experiments. Those trying to apply such strict conditions for explaining these abstract yet fundamental concepts are akin to blinds led by blinds. I only wish that more people would appreciate the value of thinking logically without the rigorous requirements of the scientific method. These scientific methods have without doubt been extremely useful to analyze the external world to a great extent. But to expect these methods to be enough for analyzing concepts like consciousness is somewhat akin to trying to "see" with ears while deliberately closing the eyes. It is not an appropriate tool for this purpose.

There are two ways of looking at the nature of consciousness. In the limited view of consciousness that we have, each person has a different consciousness. Each being thus has a distinct consciousness that is different from the consciousness of other beings. This creates the concept of "I" and "you" and can be called as ego. In this view, everyone is for himself. This creates a conflict between people and thus differences within the society. Each person wants to gain at the other's expense. As a result, nobody is happy. What we need is a more elaborate and unrestricted view of consciousness. In this view, consciousness is an all-pervading phenomenon. Each individual being is permeated by the one absolute consciousness but can sense only that part which lies in his body. As long as an individual identifies himself with this limited consciousness, there will be ego, conflict and as a result, unhappiness. As soon as one realizes that the consciousness is not piece-meal but is a part of or rather the absolute consciousness itself, the ego will vanish. There will be no conflict and eternal happiness will follow.

There have been attempts at searching intelligent life in the cosmos. Samples have been brought from Mars to see if there is any evidence of life there either now or sometime in the past. There have been many speculations of alien life, and whether there can be any communication between us and the aliens. Very recently, a 100-million dollar project has been launched to find evidence of intelligent extraterrestrial life. Speaking at the launch of this ambitious program, the famous British physicist Stephen Hawking said, "Somewhere in the cosmos, perhaps, intelligent life may be watching these lights of ours, aware of what they mean. Or do our lights wander a lifeless cosmos – unseen beacons, announcing that here, on one rock, the Universe discovered its existence. Either way, there is no bigger question. It's time to commit to finding the answer – to search for life beyond Earth. We are alive. We are intelligent. We must know." The curiosity of whether extraterrestrial life exists or not is, of course, a natural urge, and the attempts at finding it are laudable. Yet, I think the point is missed when one thinks about the universe being discovered only by the intelligent beings like you and me or some other aliens with equivalent or superior intellect. In my view, awareness and consciousness are not restricted to intelligent beings; these are not restricted even to living beings. Consciousness is an inherent property of the cosmos; the Universe is actually self-conscious and self-aware.

Just as space and time are all pervading entities, consciousness too, in my view, pervades the entire cosmos. Just as matter exists in a part of the all-pervading space, and just as events in the universe take place within a small frame of the eternal flow of time, even so the living beings exist as individual egos in the limitless expanse of eternal consciousness. Just as drops of water are to an ocean, even so are individuals to this absolute consciousness. So long as an outside covering of a container encloses these drops of water, they appear to be different from the ocean. As soon as the container is discarded and water is poured into the ocean, these drops of water become one with the ocean. All the differences between different containers (whether made of clay or gold) are dissolved.

The problem with mankind is that we identify ourselves with the containers, not even with the water contained within the container. That is, we identify ourselves with the body and not even with the limited consciousness within our body. The body, including the heart, mind and brain, belongs to me. "I" do not belong to the body. A few more evolved people would identify themselves with the consciousness within their bodies. Even this view is restricted, as we have already seen. Consciousness, as understood by them, is trapped within the body. This trapped consciousness actually is ego. Thus, such individuals identify themselves with their ego. Unless one becomes free from the artificial bondages of the body and ego, one cannot realize this universal consciousness. It is not easy. Although I am able to articulate a reasonable explanation for the water in a container and that in the ocean, the individual body with its ego and the limitless consciousness, I must confess that I am not able to realize and experience this truth. Very few manage to break this barrier of body and its individual ego, and realize consciousness in its whole entirety and glory. They realize consciousness and thus become one with the universal consciousness. Perhaps, Buddha and Christ were such people.

Science looks at external universe and explains things by observations and analysis, which are the functions of the body and mind. We have seen that this approach is not optimal for understanding consciousness. To understand consciousness, one has to look within and understand one's own self. Scientific method, alone, is thus extremely inadequate for understanding the nature of consciousness. We have to explore much beyond the limits set by these scientific methods. This discussion would have raised a number of eyebrows among scientists brought up and trained in the traditional ways of science in which observational confirmation of data and facts is the key to the understanding of any phenomenon in the world. In the method of traditional science, there is an observer who is insulated from the phenomenon he observes. However, quantum mechanics shows us the intimate interaction between the observer and the observed; that there can be no insulation between the two. The very act of observation alters the behavior of the observed phenomenon in the scientific world.

Through the interaction between the observer and the observed, quantum physics gives us a hint of the working of consciousness within the quantum world. We discussed the double slit experiment in the last chapter on quantum theory. The strange behavior of an electron in the double slit experiment is a clue to the inadequacies of the materialistic explanation of the quantum phenomena. But the physicists and scientists are tentative, skeptical and almost unwilling to make the next move; that is to study the topic of consciousness in all seriousness and sincerity. Such a discussion on consciousness cannot any longer be swept under the carpet. Closing one's eyes to reality does not make the reality fade away. Consciousness is an important player, perhaps the key to the understanding of the nature of reality.

The discussion on the double slit experiment in the preceding chapter brought us to an uncomfortable and weird conclusion that electrons (quantum entities) are somehow aware whether they are being observed and analyzed, and they change their nature accordingly. This problem can, to some extent, be solved by the "complementarity principle" as described by Niels Bohr, one of the founding fathers of the quantum theory. According to this, the wave and particle natures of the electrons are complementary properties inherent in the nature of the electron; they are not opposing principles. We see one of these complementary properties based on the kind of experiment we do or the measurements we make. The nature of our enquiry determines the facet of reality that we see in the quantum world. When we decide not to see which slit the electron passes through, it behaves as a probability wave passing through both slits, interacting with itself, and producing a pattern of diffraction on the fluorescent screen behind the slits. When we decide to see which slit the electron actually passes through, it behaves like a particle passing through one of the slits, producing a pattern on the fluorescent screen consistent with that of a particle. But how does the electron know that it is being observed, and how does the act of observation result in the collapse of probability wave into a particle?

Let us now bring "consciousness" into this discussion. We had discussed in the earlier chapters about the basic entities of the universe: space, time, matter and energy. We had also discussed how these entities are inter-related. It is time now to introduce "consciousness" as another basic entity of the universe. Just like space and time, it pervades the entire universe. It is not just in you and me and the other human beings, it is also in the plants and animals. Not just in the living beings, but consciousness in this sense is present in dust and sand, fire and water, marbles and stones, oceans and mountains, earth and moon, planets and stars, and so on. It is present in the deepest recesses of the black hole, in the burning core of the stars, and in the cold vacuum of the intergalactic space. It was present at the time of the big bang, it continues to permeate the entire universe now, and will be there when we won't be there to ask questions about the mysteries of nature. It is perhaps the elusive essence of reality that scientists and philosophers seek.

All this may seem and is indeed philosophical, but there is a deep scientific connection. The same consciousness that permeates the entire universe is present within each one of us, and it is also present in the quantum world of electrons and photons. The degree of our awareness about the consciousness is limited to our own bodies; it may well be that the universe as a whole is aware of its own consciousness in its entirety. Whether or not this is the case, and whether or not electrons are aware of the miniscule degree of consciousness that permeates it, the consciousness of the entire universe and that within the small electron is one. The individual consciousness is but a manifestation of the unitary consciousness of the universe. Any alteration in the consciousness at one place would have an impact on the consciousness at another place. So, in the context of the double-slit experiment, when the observer peeks in to see which slit the electron actually goes through, the consciousness of the observer interferes with the consciousness permeating the probability wave of the electron. This manipulation that takes place at the level of consciousness results in the collapse of the probability wave into a particle form of electron. This could perhaps explain the mystery of the observer-observed interaction in the quantum world that was causing so much heartburn among physicists.

The curious state of Schrodinger's cat

Figure 17: Schrodinger's cat: Is the cat dead, alive, or half dead and half alive? Quantum physics tells us that the cat is in a superposition of being dead and alive, both these states exist as potentia in the cat. The act of observation collapses this superposition of states into a single state of being either dead or alive.

Another paradox of quantum theory is the thought experiment of an Erwin Schrodinger, popularly known as Schrodinger's cat (*Figure 17*). Erwin Schrodinger was one of the key figures in the development of quantum theory, but he was quite uncomfortable with its rather strange and weird consequences. In this thought experiment, a cat is placed inside a cage along with a radioactive atom, a Geiger counter, and a flask of poison. Let us suppose that this radioactive atom, following the quantum rule of statistical probabilities, has a 50% chance of decaying in an hour. So, if the radioactive atom decays within that hour, the Geiger counter ticks, it will trigger a series of steps leading to a hammer breaking the flask of poison, and the poison will then kill the cat. This cage is sealed and no observations are allowed. What then happens to the cat after an hour? Intuitively, we would be inclined to think that regardless of whether we look into the cage or not, the cat is either alive if the radioactive atom has not decayed, or dead if it has (the probability of each event being 50%). However, this is not what the quantum theory concludes. The decay of the radioactive atom follows probabilistic rules in accordance with the quantum mechanics. Only when it is observed, the probability wave collapses to an actual defined event of a decay or non-decay, and only then the cat can be termed as either dead or alive. When it is not observed, both the possibilities of decay and non-decay are simultaneously superimposed, and hence the cat is both dead and alive, or is half dead and half alive at the same time. All of us have probably heard of a cat having nine lives, but such a state of being both dead and alive at the same time is quite extraordinary and bizarre even from the standpoint of a cat.

The strict interpretation of the quantum mechanics, the so-called Copenhagen interpretation, explains the cat's state as a superposition of being dead and alive; this superposition of life and death exists as potentia. It is the act of observation that collapses this weird superimposed state of the cat into a single state of either being dead or alive. The curious case of Schrodinger's cat has led to several debates and discussions, leading further to certain even more controversial propositions like the multiple world theory. This theory of multiple parallel universes was put forward by two physicists Hugh Everett and John Wheeler. According to them, both the possibilities of live and dead cat exist simultaneously; but they exist in two parallel universes. Every quantum event forces the universe to split into parallel branches so that each of the possibilities given by quantum probabilistic equations is taken care of. This seems more like an idea straight from a science fiction, but is considered seriously by many quantum physicists.

What is my take on this paradoxical case of Schrodinger's cat? Well, I think that in all the enthusiasm of discussing about probabilities and potentia, the superposition of states, and the parallel universes, the physicists have probably ignored or underestimated the cat's own ability to observe. The cat is certainly an observer of what is happening inside the cage. The act of observation by the cat would certainly lead to the collapse of the radioactive atom's superimposed states of decay and non-decay. So if the decay has occurred the cat is dead, if it hasn't the cat remains alive. It has something to do with consciousness within the cat observing the system, and interacting with the consciousness within the radioactive atom, thus forcing the collapse of probabilities into one concrete result: either a dead cat or a live cat.

While discussing about the consciousness within the observer, consciousness within the cat and that within the electron, we must not forget that it is the same, one consciousness. Later in his life, Erwin Schrodinger devoted much of his time to questions of biology and philosophy. In his short book on "Mind and Matter", in another context, he says, "There is obviously only one alternative, namely the unification of minds or consciousness. Their multiplicity is only apparent, in truth there is only one mind... Consciousness is never experienced in the plural, only in the singular." Schrodinger seems to echo what was boldly claimed by the Upanishads millennia ago. The ancient Indian philosophical treatises that were compiled into several Upanishads describe the ultimate reality of the universe as its supreme Self (consciousness); the same supreme Self is also within all. This universal Self is named as "Brahman" in the Upanishads and is said to be devoid of attributes. The entire material universe is subject to the categories of space, time and causation; but Brahman, the supreme reality, is beyond. In contrast to the material objects, Brahman is not in space but is spaceless. Brahman is not in time but is timeless. Brahman is not subject to causality but is independent of the causal chain. Kena Upanishad says, "That which cannot be expressed by speech, but by which speech is expressed – know that alone as Brahman, and not that which people here worship. That which cannot be comprehended by mind, but by which mind is comprehended -- know that alone as Brahman, and not that which people here worship. That which cannot be seen by the eyes, but by which the eyes see -- know that alone as Brahman, and not that which people here worship. That which cannot be heard by ears, but by which hearing is perceived -- know that alone as Brahman, and not that which people here worship. That which cannot be breathed by breathing, but by which breath is directed -- know that alone as Brahman, and not that which people here worship."

There may be many manifestations of consciousness, but it remains one single entity. It might again seem philosophical and metaphysical, but then it is the same thing about space and time. Space is one vast entity across the universe. We exist in space; the space too exists within us. Similarly time is one single entity ticking away ever since the universe began. Past, present and future are characteristics of time. If we examine closely, the same thing applies to all matter and energy as well. If we believe in the big bang theory, the entire matter and energy in the universe was once concentrated in the primordial fireball at the event of the big bang. All the different forms of matter and energy that we see today are simply rearrangements of that one matter-energy complex that got released with the big bang.

Einstein was perhaps correct when he said that God does not play dice. He always spoke about the hidden variable in quantum physics without which this theory remains incomplete. Perhaps consciousness is that missing link, the hidden variable that Einstein was talking about. The consciousness that pervades the entire universe also permeates the quantum world. The movement of electrons and photons is not random; the apparent randomness is probably because there is an element of consciousness that guides their movement, or provides a background to their existence and motion. These quantum particles (or waves) seem to have an element of free will that can be ascribed to the background of consciousness that pervades and permeates the entire cosmos. The act of observation alters the behavior of such quantum particles because observation requires consciousness to interpret it; this consciousness of the observer interacts with the consciousness around and within the quantum world and thus causes the alteration in their behavior.

18
Conclusion: Reinterpretation of zero

"God used beautiful mathematics in creating the world."

■ Paul Dirac

This whole book has been a logical attempt to define the nature of reality through a re-interpretation of zero. In the process, we have come across theories of origin of the universe, special and general relativity, quantum theory, and the nature of consciousness. We discussed how the Big bang theory of the origin of the universe fails to provide answers regarding the origin or the nature of the extremely dense primordial fireball in which the entire matter-energy of the universe was concentrated. It fails to answer why this highly dense fireball started expanding at a terrible pace creating space for itself and forming the universe. These questions, though often shunned by science, cannot be ignored philosophically. Science may be content with the question of "how," but the synthetic discipline of philosophy longs to answer the questions of "why." It is in this spirit that I have attempted to search for an answer to the origin of the universe. While my intention was originally confined to focusing on the origin of the universe, I realized that this cannot be an isolated search; the questions regarding the "hows and whys" of the origin of the universe are closely intertwined with the nature of reality itself. My search for the answers to the origin led me to even more exciting path towards trying to unravel the mysterious nature of reality itself.

Whatever we see and experience in this universe – the rocks and mountains, rivers and oceans, animals and plants, sun and moon, planets and stars, days and nights, rest and motion – all these can be reduced to certain basic components; these are space, time, matter and energy. Most of science deals with the behavior of matter and energy. Space and time are considered as given, a priori elements of nature. The drama of the universe is enacted by the matter and energy in the background of space and time. Laws of motion were described, gravitational force was explained, and orbits of planets could be predicted with great accuracy. This was the classical Newtonian physics before the arrival of Einstein. When Albert Einstein first burst into the scene, he dazzled the world with his theory of relativity. Special relativity forced us to abandon any preferred frame of reference, and redefined the concepts of space and time. No longer were these entities (space and time) uniform and unchanging; the length of the same object could shorten with increasing speeds (length contraction), and the clock on a fast moving object could run slower (time dilation). Space and time became interwoven as a four-dimensional space-time.

These developments notwithstanding, one crucial aspect surprisingly did not attract attention of physicists. The law of conservation (the first law of thermodynamics) is applicable to mass and energy. So, the sum total of mass and energy in any closed system would remain constant. If we consider the entire universe to be one closed system, the total mass and energy in the universe would be constant; they could only change forms and redistribute. However, space and time were considered as gift of nature, inexhaustible and self-generating.

The law of conservation did not apply to space and time. Although there was a radical change in our understanding of space and time after the advent of special and general relativity, the most basic of all laws of nature (the law of conservation) still did not include space and time within its purview. In the big-bang theory of the origin of the universe, the entire matter and energy of the universe was considered to be contained within the extremely dense primordial fireball; and although the universe has undergone significant evolution and change since the big bang, the sum total of mass and energy within it remains constant. The dense plasma of matter-energy complex within the primordial fireball has transformed and redistributed to form stars and galaxies and planets, but the sum total mass and energy has obeyed the law of conservation since the time of big bang. Yet, we see that the space within the universe has expanded tremendously since the big bang; it continues to expand and is predicted to continue expanding. Time continues to flow relentlessly without any halt.

So, whence is space generated? From what does time come into being? Why does the law of conservation not apply to space and time? These are disturbing questions, and it is perplexing to me as to why even the modern physicists have not considered these issues seriously. My answer to these questions lies in the reinterpretation of zero. As I have elaborated in the previous chapters, I have considered the sum total of all the entities in the universe to be zero. This solves two problems at once. First, it takes care of the issues about the origin of the universe. When we consider zero or nothingness to be the source of all that is in the universe, the question about the origin of nothingness becomes meaningless. Of course, we need to explain how and why did the universe spring out from nothingness. Secondly, it also resolves the problem about the law of conservation. In this broader view, the law of conservation applies to the entire universe, and is not just limited to mass and energy. In this view, the sum total of everything in the universe is zero; and this sum of zero would always remain constant. Zero may take different forms and shapes; it manifests in this universe as space, time, matter, energy, consciousness and possibly other things that we are unable to perceive. So long as the sum total of all these things remains zero, the law of conservation would hold in the widest possible sense.

The totality of the universe being zero and law of conservation having its widest application is fine, but this has to answer some more questions. All these entities, viz space, time, matter and energy seem so different from one another; how can they be added together? After all, addition and subtraction can be made only between like things. The weight of two objects, one weighing 1 kilogram and another weighing 4 kilogram can be added together to give a value of 5 kilogram. Similarly a distance of 3 kilometers and another of 6 kilometers can be added to give a total distance of 9 kilometers. A duration of 2 hours can be added to another duration of 6 hours to give a total time duration of 8 hours. But how can we add a mass of 3 kilogram to a time duration of 6 hours? How can we add a distance of 7 kilometers to a time duration of 10 seconds? It doesn't seem to make sense; it appears so absurd and actually seems ridiculous to add things that are so different from one another. How do we get across this seemingly insurmountable problem?

The answer to this riddle was hinted partly by the Einstein in his theory of relativity wherein he concluded that space and time are not actually different entities, but are intertwined as a four-dimensional space-time. He also established the unity of matter and energy through his famous equation of mass-energy equivalence ($E = mc^2$). The four seemingly different entities (space, time, matter and energy) are now reduced to two basic entities (space-time and matter-energy). It was Einstein once again who showed the close interaction between space-time and matter-energy. In the explanation of gravity by the general relativity, he proposed that matter-energy warps the fabric of space time. This distortion of space-time curvature by matter-energy is the cause of gravity, and a planet for instance would simply follow the curvature of space-time around a particular star. This geometrical interpretation of the force of gravity underscores the close relationship between space-time and matter-energy. It would take another bit of imagination to take a conceptual leap and propose now that space-time and matter-energy are not different entities; rather there is a common thread that links them together. To me, this common substratum that gives rise to space-time and matter-energy must therefore necessarily be zero or nothingness. The sum total of all that exists in the universe must be zero, and this nothingness or zero transforms into the myriad forms as space, time, matter and energy. The birth of space-time from zero must coincide with the origin of matter-energy and other entities and phenomena; the sum total of all these being zero. Perhaps matter-energy can be viewed as knots within space-time. The formation of knots within space-time results in the alteration of geometry of the space-time fabric, resulting in gravity.

We also discussed in some detail about the nature being discrete and granular. A critical examination of the example of overtaking vehicles and the paradoxes of the Greek philosopher Zeno led us to the conclusion that continuum is not possible in nature. This granularity of the nature of reality is the reason behind the formation of stars and galaxies. Let us understand how it is so. The discreteness of space and time allows for small differences in the density between regions in a given space. Thus, in the early universe after the big-bang, the distribution of matter and energy was not uniform. When the universe expanded, the distribution of matter became even less uniform; there were regions that had no matter or very less amount of matter, and there were regions that had more. The gravitational attraction between matters caused them to condense together, creating further discrepancies in the density of regions within the universe. This gravitational clumping of matter was the reason for the formation of galaxies and stars.

If space were continuous, the matter would have been distributed uniformly throughout the early universe after the big-bang. With the expansion of universe, the matter contained in it would also have been distributed evenly so that the density of all regions of the universe decrease uniformly and remain equal to one another. The expansion of the universe would not result in an unequal distribution of matter. The gravitational attraction between the particles in the universe would not result in their clumping or condensing together because this attraction would be the same everywhere, and would thus even out. Formation of galaxies, stars, planets, trees and humans would not be possible.

Quantum theory introduces an element of statistical uncertainty to the description of reality by classical physics. The rules of probability govern the quantum world, and nothing seems certain. A particle can also be a wave, and vice-versa. The two-slit experiment that we discussed in the chapter of quantum theory blurs the distinction between the observer and the observed. The classical concept of an external world independent of the existence of an observer is shattered. An electron or a photon seems to know when it is being observed; the act of observation changes the behavior of the electron or photon from a wave to a particle. From the quantum theory, we get a hint of consciousness interacting with the universe. To the basic entities of space-time and matter-energy that make the material universe, we can now add consciousness that permeates the entire universe. Consciousness is the creative principle of the universe and an essential component for its existence. We should now conclude that the sum total of all the entities in the universe – space, time, matter, energy, and consciousness that permeates all these – should be zero.

But why should zero transform into the entities that make the universe? Why can't zero or nothingness remain as it is? My answer to this is the hypothesis that I had proposed in the earlier chapters, that the instability of an entity is directly proportional to its proximity to zero. There is so much uncertainty and instability in the quantum world; the quantum particles/ waves are extremely minute and their proximity to zero results in their instability. The corollary to this hypothesis of increasing instability with greater proximity to zero is that nothingness or zero is infinitely unstable. This means that it is impossible for zero or nothingness to exist as such. But our quest towards the origin of the universe had led us to nothingness as the source of the creation of the universe. The sum total of the universe must be zero; yet zero cannot exist on its own. Zero or nothingness, therefore, exists as myriad of entities and phenomena the sum total of which remains zero. These entities and phenomena would always follow the law of conservation not individually, but as a whole. The universe exists because nothingness cannot.

Therefore, the universe can be thought of in two ways. In totality, the universe is zero (nothing) and we can refer to it as a formless universe. When the universe takes the form as we see it, it is a multitude of entities, the positives and negatives, so long as their sum total is zero. We, as the blind men trying to make sense of the elephant, understand the reality of the universe only selectively. Nothingness or zero, to me, is the secret to its comprehensive understanding. It is this interpretation of the nothingness (zero) that can explain the existence of the universe. Perhaps it was this seed that Ashok da, our teacher, inadvertently sowed in my mind when he told us the imaginary tale of two snakes eating each other and disappearing into nothingness. That seed has taken a form of a small book that you just finished reading.

Durgatosh Pandey is a well-known cancer surgeon based in New Delhi. He studied medicine and did his postgraduation in surgery in Delhi. Subsequently, he trained to be a surgical oncologist at Chennai and Mumbai, and received fellowships from Singapore and the USA.

He was on the faculty in Banaras Hindu University and the All India Institute of Medical Sciences (New Delhi) before embarking on private practice. Despite his busy practice, he has a keen interest in science, philosophy and mythology. "A peep into void" is his maiden book, in which he has attempted to find answers to his own curiosity about the world around us.

He lives in New Delhi with his mother, wife and two children. He loves reading and playing with his kids.

Printed in the United States
By Bookmasters